A PENGUIN SPECIAL
FOOD ADDITIVES

Erik Millstone was born and brought up in [...] parents. He gained a physics degree from [...] gained three postgraduate degrees in philosophy. Since 1973 he has held a teaching post at the University of Sussex concerned with the social impact of science and technology. For eleven years he has been studying the causes and consequences of technical change in the British food industry. This book is a product of that work.

ERIK MILLSTONE
FOOD ADDITIVES

Penguin Books

Penguin Books Ltd, Harmondsworth, Middlesex, England
Viking Penguin Inc., 40 West 23rd Street, New York, New York 10010, U.S.A.
Penguin Books Australia Ltd, Ringwood, Victoria, Australia
Penguin Books Canada Ltd, 2801 John Street, Markham, Ontario, Canada L3R 1B4
Penguin Books (N.Z.) Ltd, 182–190 Wairau Road, Auckland 10, New Zealand

First published 1986

Copyright © Erik Millstone, 1986
All rights reserved

Made and printed in Great Britain by
Richard Clay (The Chaucer Press) Ltd, Bungay, Suffolk
Filmset in Monophoto Sabon by
Northumberland Press Ltd, Gateshead, Tyne and Wear

Except in the United States of America, this book is sold subject
to the condition that it shall not, by way of trade or otherwise, be lent,
re-sold, hired out, or otherwise circulated without the
publisher's prior consent in any form of binding or cover other than
that in which it is published and without a similar condition
including this condition being imposed on the subsequent purchaser

Contents

Introduction   9

1 Adding Up the Cost of Food   11

2 What are the Food Additives?   32
    1 Consumer protectors   34
    2 Shelf-life extenders   39
    3 Cosmetics   40
    4 Processing aids   54
    5 Nutritional additions   55
    The cost and benefits of additives   55

3 Who Controls the Chemicals and How?   58

4 How are Additives Tested for Safety?   74
    The science of toxicology   74
    Human epidemiology   75
    Short-term tests   77
    Animal tests   81
    The acceptable daily intake   86

5	Who Is Being Protected?	105
	Acute hazards	106
	Hyperactivity	107
	Other kinds of acute intolerance	110
	Chronic hazards	114
6	What Is To Be Done?	129
	Appendix	150
	Notes and References	152
	Index	159

It is not from the benevolence of the butcher, the brewer or the baker, that we expect our dinner, but from their regard to their own interest.

   Adam Smith, *The Wealth of Nations*, Book 1, Chapter 2

Introduction

Some commentators erroneously believe that objectivity is only possible if one adopts an attitude of indifference to the character and consequences of social processes. My view is the direct opposite and I subscribe instead to Barrington Moore's statement:

In any society the dominant groups are the ones with the most to hide ... Very often, therefore, truthful analyses are bound to have a critical ring, to seem like exposure rather than objective statements, as the term is conventionally used [to denote 'mild-mannered statements in favour of the status quo']. For all students of human society, sympathy with the victims of historical processes and scepticism about the victor's claims provide essential safeguards against being taken in by the dominant mythology.[1]

This book aims to pierce the wall of orthodox rhetoric and reveal the truth about the use of food additives – and to show how their use serves or harms the differing interests of the food industry and consumers. It is my belief that at present industrial interests massively dominate the regulatory process. This balance of interests needs to be redressed, and this book should be seen as one element in a campaign to improve the standards and regulations covering the use of additives in Britain and overseas. The purpose of the book is not to ruin the appetites of readers, nor to tell anyone what to avoid or what to eat, but rather to give people a better understanding of what they are eating and to provide them with more information to help them decide what they want to eat.

My interest in this area began in 1974 when I read an article in the *New York Review of Books* called 'Death for Dinner'.[2] Its author, Daniel Zwerdling, argued that the American food industry was using chemical additives in ways which enabled them to enhance the profitability of their activities but failed to guarantee that no harm was being done to the health of consumers. Until I read that article

I had always assumed that the combined efforts of scientists and administrators working in industry and government would ensure that no chemical would be used in food if its use might damage consumer health or exploit consumers commercially. That article shattered my complacency.

Zwerdling only discussed the position in the USA, and so I decided to find out whether the situation was better or worse in the UK. It was not possible to find an immediate answer to that question because no work had been done in this field. From 1974 until now I have been researching and investigating the British and European food industries, with a particular interest in their use of food additives.

My research has made it clear that the ways in which food additives are used and regulated are profoundly unsatisfactory and require extensive reform. I do not claim that the problems concerning additives are more serious than those which arise in relation to pesticides, pharmaceutical products or environmental pollution; the issues, however, are important and have been neglected and ignored for far too long.

My thanks are due to the many people who have helped and encouraged me in my research and writing. The list includes, but is not exhausted by: John Abraham, Ray Barrell, Ernest Brown, Geoffrey Cannon, Mary Farmer, Sam Hanson, Jerry Jones, Jenny Kane, Jannet King, Ian Miles, Caroline Walker, David Wallace and Brian Wynne.

This book is dedicated to all its readers, and to the memories of Siegfried Mühlstein and Lindsey Sutcliffe.

1 | Adding Up the Cost of Food

Time and again the British food industry has claimed that consumers are receiving better value for money and a healthier diet than ever before. If this were true it would not have been either necessary or possible to write this book. There is clear evidence that there have been earlier times when we were better fed, and plenty of evidence proving that we could be better fed. The food supply in Britain is inferior in many respects to that available in other countries. Despite the oft-repeated and bland assurances of the Government and the food industry, there is a great deal wrong with that industry. Our diet, as determined by the industry which supplies it, does not serve the interests of customers as well as it could or should. The use of food additives is, furthermore, both crucial to and symptomatic of much that is wrong with our food.

It is not difficult to establish that additives are important for (at least some sections of) the food industry; it is sufficient just to read the labels on some products. The following are four lists of ingredients from some products. You are invited to guess what the products are: such prizes as there are for this quiz are entirely in the eyes of the beholders.

(a) Hydrogenated vegetable oil, starch, cheese powder, skimmed milk powder, yeast extract, flavouring, acidity regulator (E 331), sugar, salt, wheatflour, anti-caking agent (sodium aluminium silicate), emulsifiers (E 322, E 472e, E 482), flavour enhancer (monosodium glutamate), colours (E 160b, E 171), lactic acid, mustard, calcium lactate and antioxidants (E 320, E 321).

(b) Dried glucose syrup, vegetable fat, caseinates, acidity regulator (E 340), emulsifiers (E 471, E 472(e)), flavourings, colours (E 102, E 110).

(c) Food starch, maltodexdrin, salt, vegetable fat, flavourings, flavour enhancers (monosodium glutamate), gelling agent (E 412), sugar, onion powder, dried potatoes, cabbage, carrot, swede, citric acid (E 330), colour (E 102), preservative (E 220) and antioxidant (E 320, E321).

(d) Sugar, fat-reduced cocoa, hydrogenated vegetable oil, gelling agents (calcium sulphate, E 331, E 401, E 341), emulsifiers (E 477, E 322), lactose, caseinate, whey powder, flavourings, salt, artificial sweetener (sodium saccharin), colours (E 122, E 102, E 142, E 160a), antioxidant (E 320).

These are, in turn, the ingredients of Bisto Cheese Sauce Granules, Carnation Coffee-Mate, Batchelors Slim-a-Cup: Golden Vegetable and Bird's Chocolate Flavour Mousse. These are all highly processed products which are inextricably dependent on additives. Large parts of the food-processing industry rely substantially on additives. A central objective of this book is to explain how and why additives are used, and to assess the balance of consequent benefits and risks.

The argument will be that the bulk of the use of food additives does not serve the interests of consumers, and that our health as well as our social and economic well-being would benefit if there was a substantial reduction in the use of additives. It is the law of this land that food additives should only be used if they are necessary, if we are certain that they are safe and if their use does not mislead consumers about the quality of their food. As shall be seen, none of those conditions are satisfied. Additives are rarely necessary, hardly ever can we be certain that they are safe, and their presence frequently misleads consumers. Quite simply, the law is not being applied. Yet the government has the primary responsibility for ensuring that the food-processing industry changes its practices to observe the letter and spirit of the law. It is not enough just to have a law on the statute book, it must be applied and enforced. Governments only act, however, when organized pressure from consumer interests impels them to do so.

A wise, but rarely quoted proverb reminds us that 'a man with food may have many problems, but a man with no food has but one problem'. This book argues that the use of food additives poses numerous problems for consumers, but the proverb helps us to keep a sense of proportion. The problems of famine and chronic mass hunger in under-developed countries are vastly more important than are the problems concerning the use of food additives. But, for reasons which might perhaps be found surprising, there are connections between hunger in the Third World and the use of food additives in the industrialized countries. If we can understand these connections it can help us to appreciate both why many people starve and also why the food industries of the industrialized countries rely so heavily on additives.

We consume food additives through eating processed food products, which often contain additives introduced by the processing industry. Though not all processed foods contain additives, those that do could contain far fewer, and it is true to say that, the more processed foods you eat the greater your intake of additives is likely to be. The growth in our consumption of processed foods, therefore, has brought about a rapid rise in the rate at which we have been consuming additives. To understand why additives are used, in increasing quantities, we need to know why and how the food-processing industry uses additives and how this is tied in with its expansion. There are quite a few explanations for the growth of the food-processing industry, but most of them tell only part of the story. The standard approach reflects industrial interests in claiming that the industry's growth is primarily due to consumer demand. This approach is miserably incomplete. It is inadequate for at least two reasons: first it oversimplifies the nature of consumer demand and, second, it ignores – an important factor – the state of agricultural supplies. The actions of the food industry have more to do with the food industry's desire to increase its profits, and with the state of the farming industry, than with consumer demand.

The food-processing industry has to be seen as part of the total food system or, as it is commonly known, the food chain. The chain theoretically begins at the farm, although agriculture is now itself dependent on manufacturers of agricultural machinery and chemicals. Only very rarely does food go directly from farmers to consumers, usually passing to the latter through the hands of the food-processing and distribution system. The distribution system includes both wholesale and retail components, who buy in from food manufacturers and sell on to consumers. It is helpful to locate the processing industry in the food chain as it makes it easier to see how the industry is influenced by agriculture than if the industry is seen in isolation.

The food-processing industry buys the vast bulk of agricultural commodities, and transforms them into a wide range of processed food products, which it then sells into the distribution system, eventually to reach consumers. Before the industrial revolution 200 years ago there was no processing industry worth speaking of, but now more than 70 per cent of what we spend on food buys the processed variety.

The food-processing industry is, like the curate's egg, good in parts. Some processed food products such as wholemeal bread can be entirely desirable, while others are profoundly undesirable. There is no single or simple criterion with which to distinguish the good from the bad. The issue is not, however, whether processing is desirable or not, but when does it serve the consumer's interest and when does it occur in spite of the consumer's interest?

Food processing is a vast and dynamic industry. In Britain in 1984 nearly 10 per cent of all manufacturing output came from the food-processing industry, and it accounts for approximately 10 per cent of all manufacturing employment. This is despite the fact that between 1974 and 1982, employment in the industry declined by 14 per cent, at a time when sales grew slowly.

How effective is the farming community at providing an adequate supply of food? When we look at European farming we know from the scandals surrounding the Common Agricultural Policy that Europe's farmers are massively overproducing, creating mountains of food for which there are no markets. At a global level however, when so many people are starving, we might readily be led to conclude that there is an overall shortage of food. In fact, the reality is that even at a global level there is no aggregate scarcity of food, although tragically there are some severe localized shortages and people are starving to death. People starve despite the fact that globally there is more than enough food in the world for everyone.

Until about ten years ago it was generally assumed that people starved because there was an aggregate scarcity of food, and therefore that the answer to the problem of hunger was to increase the amount of food being produced. Since the Rome Food Conference of 1974 this view has no longer been tenable. Evidence was then presented, and has since been augmented, to show that more than enough food is being produced every year to feed every single human being. People starve because they do not own any land on which to grow food, and/or because they are too poor to buy the food which is available. Hunger is caused essentially by poverty and not by scarcity. Of course natural disasters do occur, as we are seeing in Saharan Africa, but even there the problems are as much economic and social as they are technical or climatic. It is only very rarely that the surpluses of foods go to the poor and the hungry, more often they go to those of us who are neither poor nor hungry. This is because the agricultural surpluses are bought by the food-processing industries for whom they are raw materials, and resold to us, the relatively affluent consumers of the industrialized countries.

It may come as a surprise to many people to discover that there is at present and in the foreseeable future no real scarcity, and this will probably be because they are under the influence of the ideas of the classical economist Thomas Malthus. Malthus argued in a simple but plausible way that unless people restrain their sexual impulses, the tendency in the long run would always be for human populations to grow more rapidly than their food supplies. An inevitable consequence of this would be widespread scarcity and hunger. It is surely comforting to know that the facts have not yet turned out as anticipated! Despite our Malthusian expectations, the fact is that the impact of science and technology in agriculture, for at least 200 years, has been to raise the productivity of farm work more rapidly than populations have grown. The productivity of farm work has consistently risen more rapidly than productivity in the rest of manufacturing industry, with the possible exception of micro-electronics in the 1980s.

Over the long term, the productivity of farming and the total supply of food has grown, particularly in the advanced industrialized countries. There have been intermittent and severe short-term fluctuations in both supplies and prices, but on average supplies have increased and prices have fallen, yet the demand for food has not kept pace with the supply. The European and world populations have grown, but less rapidly than the supply of foods, and the amounts of food eaten by most groups have also not risen as rapidly as supplies. Left to itself, the net result of this process would be that food prices would decline both relatively and absolutely, and consequently city dwellers would obtain their food progressively cheaper while the farming communities became generally impoverished.

As a first approximation, this is an important part of what has happened in many parts of the world. Overall, the terms of trade between the industrialized countries and the under-developed countries have deteriorated so far as the latter are concerned. It requires increasing amounts of Third World agricultural raw materials to buy a given amount of advanced industrial products. Within Britain, between the Repeal of the Corn Laws in 1846 and the crisis of the 1930s, on average the cities became richer while the rural communities became poorer. When urban incomes remained stable, rural incomes declined in absolute terms. The Great Depression of the 1930s marks the turning point in the politics of food. In Britain, the USA and much of Europe, the political costs of the social consequences of rural decline became so severe that governments decided to intervene to try to manage the production and prices of foods, and to support farm incomes. This pattern of thoroughgoing intervention in food markets was consolidated during the Second World War, and since that time all governments of all the industrialized countries have tried to manage farming and the food markets.

Before Britain joined the Common Market the UK had quite a sensible way of supporting farm incomes. It was called the 'deficit payments' system. Under the old British system, consumers used to buy food at the market price, and the government gave farmers a subsidy on the food which they sold, by making up the difference between the market price and some higher pre-set price, drawing on general taxation. Currently, however, farmers are paid on every tonne of food they produce whether or not anyone wants to buy it.

A direct consequence is that farmers have an irresistible incentive to overproduce, and as a result there are massive surpluses of food. They are producing too much food in at least two senses: first there is more food than the farmers are able to sell and, second, there is more food than is good for us to eat. Many of our diet-related health problems are a consequence of eating too much food and too many food additives, rather than too little or too few.

Although the Common Agricultural Policy of the EEC is one of the classic monuments to human stupidity, it would none the less be a mistake to assume that the EEC is always in the wrong. When it comes to the regulation and control of food additives the policies of the EEC are significantly better than those of the UK government. Indeed the view in Brussels is that the British have the weakest system of additive regulation, with industry and government in Britain trying to drag European standards down to their level rather than using European standards as an example to be imitated.

It has already been said that farmers are overproducing food; it should be stressed, however, that overproduction is a relative, not an absolute, concept. Whether any particular level of production is excessive or not is a function of the level of consumer demand. The problem for farmers is that while they have been raising productivity and the extent of their production, demand has not risen to match.

To help us understand why this is so, economists have introduced a useful term – the 'elasticity' of demand. This concept can most readily be explained using a relevant example; thus if our income were to double or if the price of records were halved, most of us who buy records would cheerfully go out and buy twice as many albums. We can say that the demand for records is 'elastic' in response to changes in both prices and incomes. By contrast, if our income doubled or the price of potatoes were halved, few if any of us would eat twice as many potatoes. We can characterize this by saying that the demand for potatoes is price and income 'inelastic'.

The demand for staple foods is particularly inelastic. As the scale of agricultural production has grown, the rate at which people choose to buy and eat basic foods has not grown. Population size has grown, but even so not at a rate to match the growth in farm output. Consequently, if our eating habits had remained constant while farmers' production grew, then the price of food would have fallen, and with it the incomes and profits of the farmers and all the firms in the food chain.

Farmers do not benefit from low food prices, and so they take many steps to try to raise prices and to reduce or eliminate surpluses, and if possible to create artificial scarcities.[1] Governments also have an important role to play in the process of creating effective scarcities, and this is the key to understanding the character and impact of the Common Agricultural Policy. As far as the food industry is concerned, surplus production and low food prices are not an unmixed blessing to them either. Food companies are quite happy to obtain their supplies at progressively lower prices, but if they have to sell their products correspondingly cheaply then their incomes and profits are likely to decline just as farm incomes used to fall before governments initiated their subsidy programmes.

There is, however, one major way out for the food industry. If they can buy cheaply but sell high-priced products then their incomes may rise rather than fall. The processing industry therefore takes advantage of low-priced agricultural ingredients and transforms them into relatively more expensive and profitable processed foods. They are able to do this in spite of the fact that the demand for staple foods is notoriously inelastic, because they can stimulate a market for non-staple foods.

The demand for non-staple food products is distinctly elastic, especially from the inhabitants of affluent urban society. Over the past 200 years the long-term trend has been for average incomes to rise substantially. When incomes rise while prices fall then a given diet can be bought for progressively less money. In conditions of rising incomes and declining prices many people have chosen to 'trade-up', in the sense that they are willing and even eager to eat larger, richer and more expensive diets. It is far easier to convince people to buy many more packets of crisps than it is to get them to eat more plain boiled potatoes. The food industry therefore takes advantage of falling raw food prices, but does not have to endure lower final consumer prices, because they can transform cheap raw foods into highly profitable processed food products.

It is important to appreciate the ways in which the contemporary role of the food-processing industry differs from its original role. The food-processing industry has its historical roots in preserving seasonal agricultural products, and making them available throughout the year rather than just at harvest time. Now, however, it is not that the processing industry picks up whatever the farmers have to spare, but rather the processing sector has come to dominate farming. Twenty years ago British farmers sold about one fifth of their produce to food processors, ten years ago the proportion had risen to about one half, and currently food processors are buying about two thirds of the food produced. Often it is the processors who have first refusal on the farmers' produce, and the retail fresh-food market has become secondary. In some cases, processors have made long-term contracts with farmers. Under those conditions it is the processor who specifies the variety to be produced, and the methods of production, and it is the industrialist who decides when to harvest or to slaughter.

Sometimes the food-processing industry even tries to become relatively independent of farming by turning to food additive manufacturers to provide them with synthetic substitutes which are cheaper and more predictable than natural foods. (See plates on p. 43).

There can be no doubt that there is an appreciable demand for processed foods, although the patterns of demand are not, as many in the industry wish to claim, sufficient to explain the actions of the industry. There are several reasons why there is a demand for processed foods and why that demand has grown. Processed foods are often promoted and advertised as products which are more convenient than the raw foods which they are intended to displace. Convenience is unproblematically desirable. Buying and preparing food can be a time-consuming activity, and in our high-pressure society time is at a premium. The food-processing industry cannot actually prove that labour is genuinely being saved by the use of its products; if you can afford wholesome products which genuinely save you labour then it is obviously sensible to choose them.

For many years, the food-processing industry has not been content merely to preserve agricultural products or to cater to fixed diets. On the contrary it has enthusiastically chosen to persuade customers to change their eating habits, and to rely less and less on unprocessed staple foods. The industry has succeeded in persuading consumers to supplement their traditional diets, or better still increasingly to replace them, with processed food products.

The food-processing industry is in the happy position of receiving a plentiful supply of raw materials and having markets ready, able and willing to buy wholesome products. What the food-processing industry does in response to these conditions, and in its search for profits, is to take cheap and plentiful fresh foods and synthetic ingredients and transform them into 'value-added', and so higher-priced and more profitable, food products. This is accomplished by subjecting them to industrial processing and the addition of chemical food additives.

The term 'value-added' is crucial and potentially misleading. It is misleading if it is interpreted to mean that consumers are getting better value for money. What it means is that the manufacturers can sell their products at higher prices than would otherwise be possible.

The products are therefore of greater economic value to the industry, but not necessarily of greater food value to consumers. Even that is too generous a way of putting it. In fact food products with high added economic value tend to be those which are nutritionally inferior in many respects to the cheap fresh foods which they displace.

As one commentator has put it: 'You can make a fortune in convenience foods.'[2] To accomplish this you should firstly develop (at least what passes for) a new product. Secondly you have to promote it to the retail sector to persuade them to stock it, and then you need to advertise to the consumers and persuade them to buy it, and to buy it repeatedly.

The example of potatoes and their transformation into crisps illustrates many of the points at issue. When we buy a standard 25-gramme packet of crisps we are buying one penny's worth of potato solids which have been processed into a cooked, preserved, flavoured, packaged, advertised and delivered potato product for 12p. Additives are invaluable for the food industry in this project. Antioxidants prevent the oil in which crisps have been fried from turning rancid, and flavourings can create the illusion that bacon, beef, prawns or even hedgehogs have contributed to their final flavour. It is of course cheaper to create the illusion that prawns have been used than to use real prawns.

While some of us think that rates of profit on that scale might be sufficient, this view would not be shared in the industry. In the early 1980s a major British food conglomerate developed and marketed an alternative potato-based snack product from which even more money could be made. These consisted of compressed and flavoured pieces of potato flour which, when fried, swell up to a large volume giving the impression that they contain far more bulk than they do. The advertising literature recommended to catering establishments that they should substitute these potato crackers for chips or French fries, because instead of providing a 180 per cent gross margin over cost from a 6-ounce serving of chips bought for 10p and sold for 28p, the new product would provide a gross margin of 300 per cent using servings weighing only one ounce, and bought for 5p to be sold for 20p!

In 1985 British consumers were expected to buy some 3.8 billion packets of crisps at a cost of £467 million. The most authoritative estimate of the total scale of value-adding in the British food industry states that in 1979 British food manufacturers bought £3100 million worth of food from UK farmers and £2900 million worth of food from overseas producers, and sold approximately £12,000 million worth of products.[3] This means that the food-processing firms have just about doubled the value of what to them are just raw materials. I am not suggesting that there should not be a food-processing industry, nor an effective modern distribution system, but I am trying to point to some of the problems which arise for consumers as a consequence of the fact that the processing and distribution industries exist primarily for their own self-enrichment rather than to nourish their customers.

In order to succeed in the food-processing industry it is vital to give your product some distinctive characteristic that will enable it to stand out against the competition. You can give it a special flavour, or colour, or texture, or you can promote a distinctive brand image. Of course, if you can do all of those simultaneously then you may well be on to a winner. The food-processing industry has devoted a great deal of energy to devising new products, new ways of making old products, and new products which give the impression that they are just like some of the old products. The food industry calls this n.p.d. (new product development).

Until 1982 the British government was content to allow the food-processing industry a free hand, subject only to minimal safety regulations. In September of that year, however, the Cabinet received a report on the food industry from the highly influential Advisory Council for Applied Research and Development.[4] They recommended that the government should actively encourage and subsidize research and innovation in the food-processing industry and not solely concentrate on agricultural research and food safety. The committee which prepared the report included senior figures from companies such as Unilever and Rowntree Mackintosh and the directors of the food industry's research associations. It was hardly surprising that such a committee recommended that the government should give their industry lots of money. What was slightly more surprising, though, was the enthusiasm with which the Government greeted the report.

In 1985 the UK food industry will probably have spent a bit more than £90 million on product research and development in pursuit of expanded commercial opportunities. This goal is reflected in the character of the research which gets pursued. My favourite example concerns the occasion during the early 1970s when a leading American confectionery manufacturer obtained a state grant to pay a university researcher to breed a smaller almond nut. This was required by the company because at that time of high rates of price inflation, they preferred to produce smaller bars rather than to raise the price. They wanted a smaller nut so that as the size of the bars shrank it would not show.

If you want to make your fortune in the food-processing industry it is not enough to produce an attractive product, or even a wholesome product, you have to promote and advertise it heavily. This is because consumers do not spontaneously and autonomously desire or demand the products which the process industry wishes to sell.

The current pattern of demand for foods has not arisen spontaneously from the needs and wishes of consumers. If it had, then the food industry would not have spent in 1985 more than £300 million on advertising and promoting its products. In 1984, seven out of the ten companies which spent most on television advertising in Britain were selling food. Some parts of the food industry advertise far more heavily than others, but at a good first approximation the scale of advertising expenditures are in proportion to how highly the product has been processed. Nobody tries to advertise raw wheat or corn, but breakfast cereal products, for example, are heavily promoted. It is also possible to advertise a branded product in ways which are not possible with staple foods. You cannot advertise the potato as you can a potato crisp.

The introduction of branded products is one of the primary devices which food manufacturing firms have adopted to support their promotional activities. It is difficult to advertise and profit from a generic product in the way that you can from a branded product. Additionally, the sheer diversity of brand names can serve to seduce consumers into believing that the food industry consists of many relatively modest-sized firms rather than being dominated by a few massive corporations. This illusion is belied, for example, when we discover that a conglomerate such as Unilever includes such brand names as Birds Eye, Walls, Batchelors, John West, Liptons, Brooke Bond, MacFisheries and Mattessons – just one example of the intense concentration which characterizes this industry, illustrating the point that the most successful and profitable parts of the food chain are operated by massive industrial corporations.

McCain, a company which produces frozen chips, estimates that they spend almost as much on advertising the product as they do on wages, while Kelloggs spends as much on advertising as on wages. It is worth appreciating that Nestlé's advertising budget is far greater than the total annual budget of the World Health Organization. If you speak to the marketing people in the food industry they will readily admit that they are trying to stimulate rather than merely respond to a market.

In recent years the intensity of competition between the giant firms of the food industry has been hotting up. This has been reflected in two major changes: there has been an increasing tendency towards mergers, takeovers and intensified concentration of ownership, and there have been increases in the rates of expenditure on food product advertising. In early 1985 it was reported that Nestlé, Unilever, Kelloggs and General Foods were all intending to increase their advertising budgets by between 20 and 25 per cent. In the year to August 1984, the amount spent just to advertise snack foods in Britain totalled some £17.5 million. While the food industry as a whole spends no less than £330 million a year on advertising, the total annual budget of the Health Education Council to promote healthy eating was less than £500,000. That is less than one sixth of one per cent of the industry's advertising costs.

One group to whom advertisements are particularly directed is children. There is plenty of evidence to show that they are particularly vulnerable to commercial messages, and that this kind of advertising is extremely successful and cost effective from the industry's point of view.[5] There is also plenty of evidence to show that working-class families are far more likely to consume a higher proportion of processed foods than the middle or upper classes.[6] The consumption of cola drinks is highest amongst some of the poorest children. In general, the cheapest and most nutritious foods are the least heavily advertised, while the products most heavily promoted are often nutritionally the least desirable.

The British food industry is not only busy trying to maximize their sales, they even complain that the British are not sufficiently willing to buy more of their products. The magazine *The Economist* actually contained an article which complained that the British stubbornly refuse to eat dearer food as they get slowly richer.[7] I also recall a conversation with a salesman from the confectionery company Rowntree Mackintosh during which we observed that the further north you travel in Britain the higher the sugar consumption you find. To which he replied, 'Yes, but you can't get the buggers to eat more than two pounds a week.'

The food-processing industry, therefore, has succeeded in reducing (though not eliminating) some of the surpluses of farm products. In so doing they have benefited the farmers and the government, and as far as their own interests are concerned they have increased the aggregate value of the food market far beyond what would otherwise have been possible. One estimate is that processing and distribution costs currently account for as much as 56 per cent of consumers' total expenditure on food.

The market for processed foods has shown itself capable of very rapid growth. For example, in 1973 the UK snack-food market sold 3725 million packs of savoury snacks, but by 1983 that figure had risen to 5190 million. In the last ten years the market for crisps alone has seen a volume growth of 40 per cent. By 1983 the UK savoury-snack-food market was worth some £663 million, which is more than twice the size of the markets for tea, instant coffee or breakfast cereals. The total snack-food market had reached £3244 million by 1982 out of a total food market of £23,550 million. The confectionery bar sold under the name of 'Milky Way' used to be advertised as a snack which you can eat between meals, but which does not ruin your appetite. That was an extraordinarily revealing advertisement. Whatever it might do for consumers it gives the industry precisely what it wants. It is a cheap and profitable product which supplements your consumption and does not displace any sales. For the food industry that represents a licence to print money.

One of the more extraordinary features of the rapid expansion of the market for processed foods has been the almost continual chorus of complaints from within the industry that they are not making enough money, that their rate of profit is too low. The food manufacturers are for ever complaining that they are being squeezed between the subsidized farmers and the concentrated power of the retail chains. While we can readily accept that farmers are over-subsidized and that the retail sector is highly concentrated, we should not make the mistake of accepting the manufacturers' complaints at face value. The evidence is that their gross and net profit margins compare perfectly adequately with other components in the food chain, and other parts of manufacturing industry. There have in fact been periods when the food manufacturing, and food-additive-manufacturing industries have done exceptionally well for themselves.[8]

The main strategy of the food industry has therefore been shown to be to increase the value-added component of the foods which we buy and eat, and food additives have a vital role to play in this process. Food additives are used by the food industry to perform a very wide variety of technical functions, but what they all have in common is that they are chemicals which are deliberately introduced into processed-food products, and they are used only if they contribute to the addition of commercial value and making the products easier to sell. This is clear from the advice given in the journal *Food Engineering* which advises American food companies to '... shy away from price-oriented commodity items and look to highly manufactured products in the decade ahead. The more additive-addicted [sic] foods [that are] created the higher will be the profit margin.'[9] The validity of this account of the operations of the food industry and the role of additives can be seen reflected in the ingredients and composition of many products on supermarket shelves. The products listed at the start of this chapter should serve to illustrate the point, and an extreme example might powerfully reinforce it.

I have yet to discover a product with a longer list of additives in Britain than a Birds Raspberry Flavour Trifle. The ingredients come under five separate headings. They are as follows:

Raspberry Flavour Jelly Crystals: Sugar, Gelling Agents (E 410, E 407, E 340, Potassium Chloride), Adipic Acid, Acidity Regulator (E 336), Flavourings, Stabilizer (E 466), Artificial Sweetener (Sodium Saccharin), Colour (E 123). Raspberry Flavour Custard Powder: Starch, Salt, Flavourings, Colours (E 124, E 122). Sponge, with Preservative (E 202), Colours (E 102, E 110). Decorations, with Colours (E 110, E 132, E 123, E 127). Trifle Topping Mix: Hydrogenated Vegetable Oil, Whey Powder, Sugar, Emulsifiers (E 477, E 322), Modified Starch, Lactose, Caseinate, Stabilizer (E 466), Flavourings, Colours (E 102, E 110, E 160a), Antioxidant (E 320).

One noteworthy feature of the product is that it contains no raspberry whatsoever but it does contain no fewer than twenty-two additives, and these provide flavour, colour and texture to what are just sugars, starches and fats. We have no way of knowing exactly how many additives there are in total, because all we know is that there are twenty-one chemicals plus anything from three to several dozen flavourings.

In summary therefore we can conclude that since customers can be persuaded to change their eating habits, and to replace low-cost fresh unprocessed foods with more highly processed and expensive products this helps to solve some of the numerous problems for both agriculture and governments. It also raises the profitability of the food industry, and the industry relies on food additives to make this possible.

All this may be true, but you still are entitled to ask: so what? If we know what the processing industry is doing, and if we know that the additives are there, and that they are safe, then there is nothing more to be said. The questions to which I shall therefore turn are: what are they being used for, how much do we know about them, and are they safe?

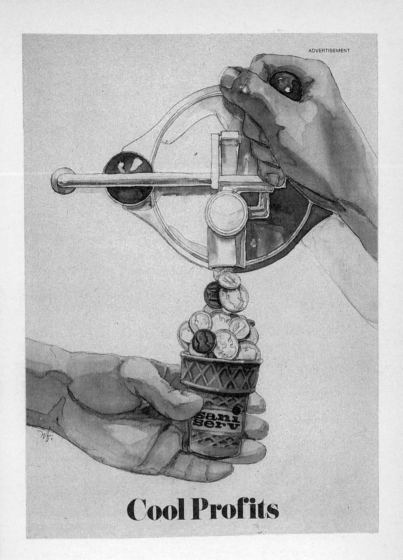

TASTE THE
EXTRA PROFITS

from automatic chain broiling with a
MARSHALL
AUTOBROIL™
... the answer to your volume feeding requirements,
regardless of your size or operation.

EXTRA PROFITS FROM:

SAVINGS

Savings in ENERGY — can shut off half of unit. Low voltage requirements.

Savings in LABOR — fewer employees needed to operate. One station operation possible. Automatic, reliable. Employees need minimum training to operate. Exclusive features assure truly automatic self-cleaning. Extra loading area for safety and increased production capacity.

Savings in REPLACEMENT COSTS of heating elements — breakthrough in design and manufacturing technology means longer life and lower replacement costs.

QUALITY

SUPERIOR FINISHED FOOD PRODUCT — units are adjustable for uniform, consistent, highest quality product 24 hours a day, 7 days a week — no peak hour variations. Automatics so that there are minimal people mistakes. Unique self-cleaning feature assures no ash residue on food.

CONSTRUCTION OF EQUIPMENT — rugged, stainless steel materials and design yield long life — heating elements are of space age alloys.

LOW PRICE

INITIAL PURCHASE PRICES are very competitive. Better amortization over longer life of equipment.

MARSHALL

Sweet'N Low — NO BITTER AFTER-TASTE SUGAR SUBSTITUTE

Pat. No. 3,625,711

PARVE A blend of nutritive and non-nutritive sweeteners

USE OF THIS PRODUCT MAY BE HAZARDOUS TO YOUR HEALTH. THIS PRODUCT CONTAINS SACCHARIN WHICH HAS BEEN DETERMINED TO CAUSE CANCER IN LABORATORY ANIMALS.

Use one packet of Sweet 'N Low for the sweetness of two teaspoons of sugar in hot and cold drinks, fruits, cereals and cooking. Each packet contains approximately 9/10th gram of carbohydrates equivalent to about 3½ calories. This should be taken into account by diabetics.

Ingredients: Nutritive dextrose, 4% sodium saccharine (40 milligrams per packet or 20 milligrams per each teaspoonful of sugar sweetening equivalency), cream of tartar, and drier

APPROXIMATE ANALYSIS: No protein. No fat. Available carbohydrates 94%.

NET WT. .035 OZ. (1 GRAM)

CUMBERLAND PACKING CORP. • BROOKLYN, N.Y. 11205

2 | What Are the Food Additives?

Many chemicals are found in our foods. Most are natural constituents of the food, some are present because of deliberate human actions, and others enter by accident. When we refer specifically to food additives, however, we are talking about a precisely defined sub-set of chemicals; indeed they are a sub-set of those chemicals which are deliberately added to food. We reserve the term 'contaminants' for those chemicals present by accident, and deal with them separately.

There are two main groups of chemicals which are deliberately added to food. The first group are the agricultural chemicals used by farmers. These include pesticides, herbicides and fungicides which are used on plants, and hormones and antibiotics which are given to animals. Although agricultural chemicals are deliberately introduced into our food supply we do not count them as food additives. They are supposed to be used in ways which ensure that they are either entirely absent by the time the food is eventually eaten, or present in only insignificant quantities. Food additives and agricultural chemicals are used by entirely different industries and they are regulated – or escape regulation – in quite different ways.

It is important to distinguish between food additives and agricultural chemicals. From the research which I have done, I am certain that there are many books begging to be written on the use of agricultural chemicals and their social impact. Although agricultural chemicals are beyond the scope of this book, many of the arguments that I shall be presenting would apply equally well to chemicals used in farming. The expression 'food additives' will be used only to refer to chemicals which are introduced deliberately into our food by the food-processing industry.

The dividing line between foods and food additives can itself be a tricky one. Strictly speaking we might want to call sugar and salt additives rather than foods, but that is not how they are defined under British, American and European laws. My usage will conform to the official distinctions, and salt and sugar will not be counted as additives.[1]

If you ask the food industry why they are using additives they will tell you that they are used in response to the demands of consumers, as shown by the fact that we choose to buy food products containing additives. For a whole number of reasons this is not a convincing explanation, and this book will aim to prove that food manufacturers choose to use additives because they serve their purposes rather than the consumer's. But what purposes does industry have which can be met by the use of additives? One well-informed apologist for the food industry gives some useful clues when he says:

... convenience foods ... usually require more additives than conventionally cooked foods because they are often [produced] under more severe conditions of temperature, pressure or agitation ... Convenience foods are generally highly processed foods, and, by even the most modern techniques, processing generally alters flavours. Although the consumer has often accepted the traditional blander flavour of canned and frozen fruits and vegetables, he now demands that his store-bought dinners approach the home-cooked variety in taste, texture and appearance ... Flavours make up the largest single category of food additives ... The addition of flavouring to a food can supply a taste where little or none existed, it can intensify, modify, or mask an existing flavour ... The growing consumption of soft drinks ... is responsible for most of the increase in the per capita consumption of food, and thereby, of food additives. Aside from water and sugar, soft drinks are made entirely of additives such as colours and flavours, non-nutritive sweeteners, thickeners and stabilizers, emulsifiers and preservatives.[2]

What this reveals is that, while processed food does not necessarily require additives, many of our existing products could not be manufactured or sold if their additives were removed. Some products, such as many soft drinks, instant desserts and snack foods, simply could not exist without additives. Some others could exist without them but would have to be produced, stored, distributed and sold quite differently. Industrial processing can be very severe on food; so severe in fact that it can damage or even destroy the colour, taste and texture. Some ingredients may have so little colour or taste, or none at all, or even tastes and colours which are unattractive, that industry chooses to enhance the appeal of their products by introducing chemical additives. As the director of technical services of a leading flavour manufacturer has put it: 'A food which is otherwise unacceptable in terms of taste or appearance can be made acceptable by the judicious use of additives in its formulation.'[3] I am not suggesting that the use of additives never serves the interests of consumers, but the interests and choices of consumers are not the dominant considerations influencing the actions of the industry.

There are many ways of dividing additives into groups; under British and European laws, no fewer than twenty-three categories are used. I shall distinguish these into five main functional groups. It should be understood that these functions are not exclusive – a particular additive may perform several functions either simultaneously or in different contexts. The five main functions which additives perform are: protecting consumers, extending the shelf-life of products, achieving a cosmetic effect, facilitating industrial processes and, finally, improving the nutritional quality of the product.

1 Consumer protectors

Some food additives can be and are used to protect the health of consumers. There are two kinds of possible risks, namely short-term and long-term, which shall be dealt with in turn.

Unfortunately good food goes bad. All foods will rot, but some do so faster than others. Meat, fish and dairy products are amongst some of the least stable foods. They are especially vulnerable to bacterial contamination. Some bacteria are harmful, and others benign, but the bacteria responsible for salmonella and botulism are particularly poisonous to humans. Indeed the expression 'food poisoning' is usually reserved for acute illness from bacterial contamination. Chemical preservatives are food additives which are used to inhibit the growth of bacteria and prevent contamination and spoilage. In Britain over recent years the number of officially reported cases of bacterial food poisoning has been rising rapidly, and everyone agrees that the number of cases which get reported is a small fraction of the real total.[4]

The levels of bacteria which are present in foods depend not just on the intrinsic properties of the foods, but also on the ways in which those foods are produced and handled. If we want to prevent bacterial food poisoning then the most important consideration is to ensure that we have good hygiene in food handling, preparation and storage. There is some (official) speculation that the increase which we can observe in the frequency of bacterial food poisoning in the UK is a consequence of our eating more meals out of the home, bought from commercial caterers. The implication is that standards of hygiene in some commercial establishments leave a lot to be desired. As a senior government scientist pointed out, in 1983 the average stay of workers in the kitchens of Heathrow airport was no more than six weeks.[5] Under those circumstances it is difficult to expect the staff to achieve high standards of cleanliness. Hygienic practice is always desirable, but not always sufficient and must often be combined with other ways of preserving food, like freezing, drying, or the addition of chemical preservatives.

There is some evidence that carcass meat from animals fed with antibiotics is particularly likely to be contaminated with bacteria, and particularly with antibiotic-resistant strains of bacteria. Most of the meat which is consumed in industrialized countries comes from animals fed on both hormones and antibiotics.

The use of what is called mechanically recovered meat also contributes to high levels of bacterial contamination in products containing it. There are several different types of machinery which can recover parts of a carcass which even the best trained and equipped butcher cannot reach. The flesh of animals and fish can be 'mechanically recovered' by taking what remains of a carcass and subjecting it to severe treatment with high pressure crushers, jets and brushes. The product from a process of mechanical recovery is a pulp or slurry which can be incorporated into such things as sausages, meat pies, burgers, fish fingers and pastes. There are many problems with mechanically recovered meat but one central issue is the fact that it is biochemically unstable and so particularly vulnerable to bacterial contamination, and therefore can only be used in combination with chemical preservatives.

During 1984 a major British manufacturer of yoghurt relaunched its product range as 'preservative free'. It was able to remove the preservatives from its products because it had invested in modern technology which improved standards of hygiene and thereby made the chemicals redundant. Examples such as this show that if there were improvements in farming and manufacturing practices there would be far less of a need for chemical preservatives in the first place.

Under British law there is an official definition of the term 'preservative' and it is quite revealing. According to the regulations, a preservative is: '... any substance which is capable of inhibiting, retarding or arresting the growth of micro-organisms or any deterioration of foods due to micro-organisms *or of masking the evidence of any such deterioration.*'[6] (Author's italics.) What this reveals is that one function which preservatives may perform is that of concealing the fact that food products are old and deteriorating. In so far as this happens, additives are being used in a way which is directly against consumer interests, because they are being used to deprive consumers of information about products to which we are entitled.

We are using quite a few different technologies to preserve foods and to prevent bacterial spoilage. We can pasteurize milk with heat, we can freeze, bottle, dehydrate and irradiate foods, and we can also add chemical preservatives. All these methods have a useful part to play. Using these technologies singly or in combinations, we know how to eliminate the risk of acute bacterial food poisoning. It is in this sense, and perhaps in this sense only, that we can say that our food supply is now better than it has ever been before.

Having explained the contribution which preservatives can make in protecting us against acute food poisoning, we come to the question of whether additives can protect us against any chronic health problems. The distinction between acute and chronic food poisoning is an important one. Acute problems are those which occur almost immediately after eating the poisonous food. The severity and the circumstances of acute food poisoning are often such as to enable us to identify the food, and sometimes even the organism, responsible. Chronic food poisoning is an entirely different matter. Chronic illnesses, such as heart disease and cancer, are problems which are not immediately apparent upon eating a meal containing foods which contribute to those illnesses. It is almost always impossible to trace the direct cause of particular cases of these illnesses. Chemical food preservatives, and other techniques for preservation, can be effective in preventing acute food poisoning, but they generally play no role in the prevention of chronic poisoning, although they may themselves contribute to causing chronic ill health. If a chemical is to be acceptable as a food preservative, then we ought to be confident that it does not itself present a chronic risk. One of the major problems with some of the chemical preservatives is that there is evidence to suggest that they may be long-term slow-acting poisons.

We know a great deal about the causes and prevention of acute food poisoning. We know very little about either the causes or prevention of chronic poisoning. Many foods naturally contain small amounts of chemicals which are known to be toxic to laboratory animals when forced upon them in large doses. There are also doses at which additives can upset the health of laboratory animals. I know, however, of only one case where the claim is made that an additive can act to protect consumers against chronic hazards from food, and this concerns the possible effects of antioxidant additives.

At varying rates, all fats and oils will combine with oxygen in ways that cause them to turn rancid. As fats and oils rancidify they become extremely unpalatable, and so antioxidant additives are used to prevent this from happening. They function primarily as shelf-life extenders, because without them so many products would have a shelf-life of a few days rather than weeks or months. Some commentators claim, however, that antioxidants can also serve to protect consumers against possible chronic risks, while others assert that some antioxidants may themselves constitute a chronic toxic hazard. There is a storm of controversy about the safety and toxicity of two of the most commonly used synthetic antioxidants, known as BHA and BHT (abbreviations for butylated hydroxyanisole and butylated hydroxytoluene; also known as E320 and E321 respectively).

There is a theoretical possibility that severely oxidized oils and fats may contain the products of rancid chemical decay, which might cause cancer, although attempts to confirm this in laboratory studies with animals have so far produced only equivocal results. In so far as rancid oils are poisonous, to that extent antioxidants should be counted as consumer protectors. Many traditional vegetable oils, such as olive oil, contain naturally occurring antioxidants which considerably reduce the rate at which they will oxidize. Most of the fats and oils which are used by the food industry would naturally break down very rapidly at the high temperatures reached during frying or baking. As a result, the vast proportion of all the oils and fats sold to industry and to consumers contain synthetic antioxidants, and this gives them the resilience which we have come to expect.

There is a good deal of controversy about the potential long-term significance of the consumption of BHA and BHT. The issue is extremely complicated, and there is nothing resembling a consensus on the matter. Antioxidants, particularly the synthetics BHA and BHT, may be chemically involved in quite a few different chronic problems, sometimes negatively and sometimes positively. Different scientists point to different possible mechanisms, some of which might do us harm, while others may protect us. Some go so far as to claim that munching antioxidant supplement pills will enable us to attain what they call 'maximum life span'.[7] Some claim that antioxidants can help to protect us against toxic hazards from suspected carcinogens by stimulating the production of enzymes in the liver, while others argue that BHA and BHT may themselves be carcinogenic to humans, since there are doses at which they cause cancer in laboratory animals.[8]

There is evidence from animal experiments that both BHA and BHT can, at various dose levels, and in different species, both inhibit and promote cancers at varying sites. We do not yet have enough information to enable us to determine whether these antioxidants are able to perform a consumer-protective function, or whether they are themselves hazardous to humans, but we can be certain that irrespective of their toxicological significance, their primary function is to extend the shelf-life of food products.[9] One supplementary function which they intermittently serve is to stabilize some other chemical additives such as colours.

2 Shelf-life extenders

One of the main functions which additives perform is to extend the shelf-life of food products. Both preservatives and antioxidants, amongst others, can perform this function. It should be noted that where a chemical additive extends the shelf-life of a product it is performing an economic rather than a consumer-protective function. Particular chemicals may perform both functions simultaneously, and even a cosmetic function too. For analytical purposes, however, it is important to distinguish between the different functions, because otherwise we might fail to appreciate how the benefits of their use are distributed between consumers and the food industry.

Antioxidants can be considered shelf-life extenders because they prevent the unpleasant flavour of rancid oils and fats, caused by decay, from occurring. Once the flavour and aroma of a product has turned rancid no one wants to buy or eat it. There are a handful of other types of additives which contribute to extending shelf-lives and these include anti-browning agents and humectants which prevent products from drying out.

When it comes to estimating the quantities of additives that are being used for these and other functions, it is difficult to provide precise figures. There are *no* publicly available figures on aggregate levels of the use and consumption of additives. When I asked those in the Ministry of Agriculture, Fisheries and Food who are responsible for supervising additives in Britain for some figures, they admitted that they had none, and I was asked unofficially to share with them any which I could find.

There have been at least two commercial surveys of the size of the market for additives, one dealing just with the UK, and another which deals with all the countries of the EEC. These reports have not been published. They are available only on a confidential commercial basis. You can buy the reports, but you cannot find them in a library. The report on the UK market was published in 1980 and cost £600 a copy, while the European Market Report was published in 1982 and then cost US$1,350. The figures which are used in this book are my own estimates, based upon these reports but supplemented with further research.

Approximately thirty-five different chemicals are permitted for use in the UK and the EEC as food preservatives. When their presence is notified on the labels of food products their E-numbers will be from E200 to E299. I estimate that currently the UK food-processing industry is using more than 1700 tonnes of preservatives annually, and is paying about £2 million for them. As for antioxidants, of which fourteen are permitted, they have E-numbers which run from E300 to the E320s, and their annual cost to industry is just over half a million pounds, which pays for about 1000 tonnes.

I also estimate that in total the UK food-processing industry is spending something like £225 millions a year (+/− £25 million), buying something in the order of 3850 different additives. It is harder to estimate the total mass of additives which industry is buying, but it is probably not less than, and probably rather more than, 220,000 tonnes a year, which is equivalent to approximately 4 kilogrammes per person per year. From these figures, it follows that together preservatives and antioxidants account for no more than 1.5 per cent by number, a few per cent by value, and less than 1 per cent by weight of the total quantities of additives in use. This means that only a very small minority of additives are used to protect consumers, or even to extend the shelf-lives of products.

3 Cosmetics

Cosmetic additives are far and away the largest group by each of the criteria of value, number and mass. I use the term 'cosmetic' to refer to additives which are introduced into foods to make them easier to sell by making them more attractive. There are two major groups of cosmetic additives. Food-industry scientists call the first group organoleptic modifiers, while the second are the textural modifiers. Organoleptic modifiers are essentially just those which affect the colour, flavour and aroma of food products, while textural modifiers affect how the products feel to our hands and mouths. The cosmetic use of additives is unquestionably the most controversial aspect of additive usage, because their use most readily benefits the food industry and their benefits for consumers are particularly questionable.

I estimate that cosmetics collectively account for something in the order of 88 per cent of the total value of the market for food additives. It is not difficult to estimate the cost to the food industry of the use of cosmetic additives, but it is impossible to estimate their costs or benefits to consumers. We can, however, demonstrate that their use is highly beneficial to the food industry. There can be no doubt that it is far cheaper to manufacture a pizza with synthetic substitute cheese and tomatoes than it would be using real cheese and real tomatoes. For example, one American advertisement for synthetic tomato flavour proclaims that: 'Cooked, ripe tomato flavour [is] offered in either dry or liquid form ... [it] is recommended for use in soups, sauces, dips, salad dressings and convenience foods ... one pound replaces the flavour and aroma of 1200–1600 pounds [of] tomato juice at a cost of $5 ...'[10] There can be no doubt that the use of cosmetics is of massive economic benefit to the food industry, but any benefits which may accrue to consumers are far harder to identify or to estimate.

If the food industry were to desist from using all cosmetic additives, but otherwise continue to manufacture their products in precisely the same way, then they would be producing a great deal of pale, tasteless, flaccid and poorly mixed sludge.

Flavourings

Flavouring additives constitute the largest single group of cosmetic additives, both in terms of the numbers used and their cost. I estimate that in 1985 the UK food-processing industry spent approximately £55 million for something like 3500 different flavourings, and that

the synthetic flavours alone account for approximately £25 million a year. Flavourings are extremely powerful chemicals which are so highly reactive that they can achieve their intended effects even at very low concentrations. In 1985 the UK food industry used about 1000 tonnes of flavouring concentrates, but this was sold in diluted forms amounting to some 45,000 tonnes. The value to the industry of these chemicals is entirely out of proportion to their cost. Despite the fact that they are so numerous and economically so significant, flavourings are effectively unregulated in the UK and the EEC. Since flavourings are not subject to specific regulations no 'E' numbers have been assigned to them, and even when their presence is indicated on labels and packets we are only told that the product contains flavourings but never which ones.

Flavouring additives can be used not just to reinforce natural flavours, but also to provide products which would otherwise be quite tasteless with distinctive flavours. They can, moreover, be used to conceal or overlay what would otherwise be unattractive flavours. In principle, the law is supposed to prevent the use of additives in ways which mislead consumers, but legislation and regulation are ineffective and consumers *are* being misled. Consider the distinction between the meaning of the two apparently similar terms 'flavour' and 'flavoured' as used on packets in the UK. A product which is labelled as 'bacon flavoured' must derive its flavour entirely or primarily from bacon, but a 'bacon flavour' product need contain no bacon whatsoever, but should merely have a flavour which resembles that of bacon. There can be no doubt, given that most people are unaware of this semantic subtlety, that this labelling device is systematically misleading, and that it is intended to mislead. When challenged on this issue the food industry plaintively laments that no one has ever thought of a better form of words; but if anyone suggests that they should use either 'artificially flavoured' or 'imitation flavour' they seem to turn a deaf ear.

It is not just traditional methods of industrial processing which destroy flavour – some of the latest processing technologies do so too, and if anything even more severely. For example, high-pressure extrusion cooking is very tough on the flavour of the product.

Extrusion is being used extensively to produce, for example, savoury biscuits and fried snack foods. As the cooking process occurs very rapidly many of the traditional flavour-generating reactions do not occur. To compensate for this, extruded products are often toasted because they would otherwise be very pale, and have flavour sprayed on to them either as a powder or dissolved in oil.

The expert committee which advises the British government on food additives has recommended that flavourings should be as fully regulated as other classes of additives. The industry has, however, mounted a discreet but entirely successful campaign to keep flavourings free from systematic regulatory control. As a result, we know very little about the flavouring chemicals which are used. Given that they are not regulated, the additive and processing industries have no incentives either to test them for safety nor to be open about their use.

Until 1985 we would have had to say that nobody even knew the identity of all flavourings used, or the quantities involved. Recently, however, the Ministry of Agriculture, Fisheries and Food, has asked the flavouring manufacturers and the food-processing industry to tell them which flavourings they are using. The government encouraged industry to co-operate in this voluntary exercise by promising that any information which firms provided would remain entirely confidential. We can now conclude, therefore, that a few civil servants do know which flavourings are being used, and in what quantities, but they are not allowed to share this information with the public.

It is not hard to understand why manufacturers wish to keep their recipes secret, but that does not mean that consumers' interests would not be better served by making all the information publicly available. The reasons for secrecy are not technical, but solely commercial. Food manufacturers do not trust each other, and fear that their competitors will steal their formulae or recipes. The irony is that while manufacturing firms do not trust each other, consumers are being expected to trust the entire industry.

Estimates of the number of flavours in use vary from as low as 2000 to as high as 20,000. My estimate, however, is that approximately 3500 flavouring substances are being used. Given that we do not know which chemicals are used, nor the quantities in which they are used, we are in no position to judge their safety.

The food industry not only makes unrestricted use of flavourings but also makes extensive use of a group of chemicals which are called 'flavour enhancers', the most famous or notorious of these being monosodium glutamate (or MSG). Flavour enhancers are chemicals which in themselves are just about tasteless, but which trick our taste buds so as to give us the impression that foods containing enhancers have more flavour than they really do. I estimate that in 1985, the UK food industry used approximately 6500 tonnes of flavour-enhancing chemicals, the bulk of which was MSG; this cost them approximately £7.5 million and was, in commercial terms, an investment that paid for itself many times over. Flavour enhancers have not yet been assigned 'E' numbers. They have, however, British numbers which may eventually be given the 'E' prefix. Monosodium glutamate is additive number 621; the other flavour enhancers have numbers between 620 and 635 inclusive.

Artificial sweeteners are a special group of flavouring additives, which for official regulatory purposes are treated separately. An employee of a company which manufactures sweeteners once suggested that artificial sweeteners should be regarded as consumer protectors because they can be used to keep people from eating sugar. Excessive consumption of sugar may indeed be bad for our teeth, the heart and the pancreas, but that does not mean that artificial sweeteners are included in foods to protect consumers. They are there for cosmetic purposes. Much of the enthusiasm which we have for sweet tastes is not natural or inevitable, but it is rather a product of the marketing strategies of the food industry.

Diabetics have particular problems metabolizing sugar, and the value of artificial sweeteners to them may arguably be greater than to the rest of us, even though they are still not necessary. The average consumer would probably be healthier and better off with reduced sugar consumption, but this does not necessitate the use of artificial sweeteners. People may wish to eat artificially sweetened products, and if safe sweeteners are available, then their use should be permitted. There are, however, many questions about the safety of sweeteners.

There has been a controversy about the safety of saccharin since 1878. In 1972 the staff of the US Food and Drug Administration (FDA) estimated that a ban on saccharin would cost the American diabetic food-processing industry something between $600 million and $1.96 billion per year. None the less, in 1977 the FDA called for an outright ban on saccharin because it is known to cause bladder cancer in rats, but the scientists were overruled by the politicians. At the time it was claimed that politicians reached their decision because of pressure from consumers. That may have been true, but the 'consumers' in question were not the American public but the soft-drink manufacturers who relied on saccharin for their 'low-calorie' product ranges. Until recently saccharin had the dominant share of the artificial-sweetener market in America, Britain and the rest of Europe. British manufacturers spend about £4.5 million per year to buy some 750 tonnes of saccharin.

From the late 1930s until 1969 another group of sweeteners, known as the cyclamates, was making inroads into the markets for both sugar and saccharin. In a set of controversial decisions, the use of cyclamates was then banned in the USA, Britain and much of Europe. At the time, *Private Eye* claimed that the research which led to the banning of cyclamates was paid for by leading companies in the sugar industry. That claim has subsequently been denied but not refuted. The current position is that there are renewed moves to return cyclamates to the permitted list of additives in Britain and the EEC.

In the autumn of 1983 permission was given for the introduction of two new artificial sweeteners into our food. One is called aspartame, and is sold under the trademark 'Nutrasweet'. The other is called acesulfame–K. There is an exceptionally sharp controversy about aspartame. The company which manufactures aspartame can provide you with an impressive dossier which is intended to show that aspartame is a thoroughly tested and an unproblematically safe chemical for almost all of us. A small proportion of the population suffers from an illness called phenylketonuria, and those with that

Preferred for taste.

Perfect for profit.

In recent tests over 300 men, women and children were invited to compare McCain French Fries with a leading continental brand of chips, and they proved their preference by voting McCain No.1 for appetising appearance, golden texture and true potato taste.

And we're voted No.1 for profit too.

Throughout the year, kitchen demonstrations of McCain French Fries versus competitors' chips consistently prove to caterers how our shorter frying times and higher yields spell bigger profits from the real McCain.

McCain French Fries. Preferred for taste – Perfect for Profit.

The star name in frozen foods.

McCain Foods (GB) Ltd., National Sales Office, Funthams Lane, Whittlesey, Nr Peterborough PE7 2PG. Phone: 0733 203051 affliction are unable to cope with one component of aspartame, and so are advised to avoid it. For the rest of us, it is supposed to be safe.

On the other hand, some independent investigators claim that there is a great deal wrong with the ways in which aspartame has been tested, and that it may well not be safe.[11] Whether or not aspartame is safe for most consumers, it remains the case that calling it 'Nutrasweet' is bound to mislead consumers into thinking that it has some significant nutritional value. As for acesulfame-K, we are in a particularly difficult position. No toxicological information whatsoever has been published in Britain on this chemical, and so we are entirely unable to assess its safety.

Colourings

In at least two respects, the contrast between colours and flavours is very striking. Although they are both cosmetic additives, colours are subject to comprehensive regulation, and we know more about the scale on which they are used than for any other group of additives. Colours are also amongst the most thoroughly tested additives, and this is because historically their use has been most heavily criticized, and numerous colour additives have been found to be toxic and have consequently been banned.

The large-scale commercial and industrial use of food colouring developed initially as a by-product of the development of dyes for textiles in the second half of the last century. In 1856 Henry Perkin, while analysing coal tar in a search for a synthetic substitute for quinine, came across a powerful mauve-coloured chemical, and this provided the basis for the rapid expansion of the coal-tar-dye industry. The centre of this industry moved from Britain to Germany in the 1880s, and companies such as the Bayerische Aniline und Soda Fabrik (BASF) devoted their considerable skills to isolating and manufacturing a wide range of dyes.

It was not long before some of the same colourings were introduced as additives into food. Colouring additives have a particular attraction for the food industry. Scientists have demonstrated why: if you present a group of people with several foods or drinks, each with the same flavour but with successively deeper colours, most people judge the strength of the flavours to be in proportion to the strength of the colours. If the test is repeated with blindfolds, then they are judged to be indistinguishable.[12]

Most foods have their most intense colour when they are fresh, and as they age their colour deteriorates. As a result, when consumers judge food products by their colour they will, in the absence of alternative information, assume that bright foods are fresh, and hence tasty and wholesome. Modern high-technology synthetic foods, such as textured spun vegetable protein and extruded snacks are just about colourless, and before one of these new products is marketed the manufacturer has to decide which colour to give it. Numerous products are sold as having, for example, fruit, meat or prawn flavours even though they contain no fruit, meat or prawns. Colours are used to sustain these commercially valuable illusions.

In the early 1970s Marks & Spencer experimented by removing colours from their canned peas, canned strawberries and strawberry and raspberry jams.[13] They reported that their sales of these products fell by 50 per cent. What can we conclude from this? It might imply that at least 50 per cent of consumers want their food brightly and artificially coloured, but only if certain conditions were fulfilled. If consumers had known, first, that Marks & Spencer's canned peas were normally highly coloured and, second, if they had known why the products had acquired a paler colour, then we would have been entitled to conclude that consumers were showing a preference for highly coloured food. Since, however, standard products were not then fully labelled, and the sudden change was left unexplained, it is not surprising that consumers may simply have assumed that the products had deteriorated.

The colours which we have come to expect in food do not simply reflect any innate knowledge of what is fresh and wholesome. Rather they are a reflection of the images given to us in brightly coloured advertisements in magazines and on television. Colourings can be added to ensure that the product matches the tones of the colour pictures of the product, rather than the food itself. Colour additives are included in food on the initiative of the food industry, and we will only know the extent to which consumers genuinely want them if and when they are in a position to give their informed consent.

It would be absurd to try to list all the different food products which contain colours. It is easier by far to identify those which do not, or should not, contain added colours. In 1979 the Food Additives and Contaminants Committee (FACC) of the Ministry of Agriculture, Fisheries and Food recommended that the government should ban the addition of colouring additives to food products which are intended for infants and young children. This advice has not been taken. Instead we live with what is described as 'a gentlemen's agreement' between the Ministry and the industry that companies will not use colour additives in baby foods. Unfortunately the government does not properly specify the age range covered by this agreement. At least three different ages are referred to in different documents and none is definitive. We have no way of knowing whether or not the gentlemen are all keeping their word or whether all food companies are run by gentlemen. We do know that with some products, such as sweets, soft drinks and snack products, marketing is deliberately aimed directly at children; we also know that these products rely heavily on additives, especially colours. It has been estimated that the average British child has consumed about half a pound of food dye by the age of 12.

Under British regulations no colourings should be added to raw or unprocessed meat, poultry, fish, fruit or vegetables; nor to tea, coffee or to condensed or dried milk. Occasionally, public analysts report that butchers have used a chemical additive on their meat to brighten the colour. The chemical of choice is sulphur dioxide applied as a sulphite powder. By its preservative action, it prevents the meat from acquiring a dark colour which would reveal its age. While this is illegal on carcass meat amd minced meat, it is permitted in hamburgers; but then hamburgers are often made from the poorer pieces of meat. Almost all butchers use pink strip lights which impart a rosy hue to their goods, but the combination of sulphur dioxide and pink lighting is supposed to be particularly effective. Trading Standards and Environmental Health Officers, along with public analysts, generally believe that the practice occurs far more frequently than comes to their notice.

Reports from countries in which oranges are grown suggest that some oranges are artificially coloured. If fruit growers were to wait until their fruit was ripe before harvesting it, it would then be over-ripe when it arrived in Europe. As a result, fruit is generally harvested under-ripe and then ripened artificially, for example using gases. Under-ripe oranges are naturally green rather than fully orange. To provide their products with a uniform skin colour, growers can inject colour dye into the trunks of the fruit trees. Most of that colouring goes into the peel, but at least some enters the flesh.

Many orange drinks are made from what are called 'concentrated comminuted oranges'; this substance is very different from what you would get from squeezing the fruit. Freshly squeezed orange juice is rather pale, and rapidly loses what little colour it originally had. If you crush whole oranges, including the peel, into a pulp, and from that extract a concentrate then you can obtain a product with a more intense and stable colour. If orange trees and orange peel are artificially coloured then the subsequent products will contain artificial colourings, even though these will not appear on the label.

Colourings can be found in many products from which most consumers would expect them to be entirely absent. Butter generally contains colour, even though colouring additives are not added directly to butter. The natural colour of butter will vary with the time of the year. Summer butter is the most yellow, while winter butter can be rather pale. The butter trade prefers a uniform colour in its products, and so yellow colouring additives are frequently introduced into dairy-cattle feed, so that it enters the cows' milk fat and gets into the butter. Because the colour is not added directly to the butter, this practice does not technically count as using a food additive and so no food additives are listed as ingredients on the label.

In the mid-1960s many consumers complained that eggs laid by battery hens had pale shells and yolks. As a response, farmers started to introduce colour additives into chicken feeds. We now have browner shells and darker yolks; in other words the eggs look as though they come from free-range hens, even though they are battery products. Apparently it has also become common practice for fish farmers to feed their trout and salmon with red dye because this gives the flesh a darker colour which untutored consumers seem to think superior.

The United States was one of the first countries to insist that colouring additives in food should be regulated. In the first decade of this century the redoubtable Harvey Wiley, then head of the Bureau of Chemistry at the United States Department of Agriculture (USDA), led a pressure group which forced Congress to enact legislation to control colours, and pressed the USDA to enforce those laws. Many European countries too have been regulating colours for most of this century. For instance, Norway has regulated food colours since 1935 and since 1980 all synthetic food colours have been banned in that country.

The use of colourings in the UK has only been subject to specific regulation since 1957, and here, as so often, the British trailed far behind the example set by other industrialized countries. In 1955 about eighty different synthetic colours were being used in food and drinks. The 1957 regulations permitted only thirty synthetic dyes, and the permitted list shrank to twenty-five in 1966, twenty-three in 1973, and in 1985 British regulations permitted the use of seventeen synthetic colourings, as well as approximately thirty colours of natural origin.

Due to the legal structure of the EEC, any colouring additive which is permitted in any one country has to be permitted in all other countries. Each state, however, has the right to decide the quantities and the products in which those colours may be used. As the British food industry insists on using a red dye called amaranth (or E 123) which France and Italy wish to ban, these countries meet their community obligation by permitting it only in microscopic quantities in caviar.

Because colouring additives are so controversial, they have been subject to particular scrutiny, both by governments and by toxicologists, and they are the only group of chemicals for which we have any published official figures on consumption rates. The 1979 Report of the Food Additives and Contaminants Committee contains a great deal of data, but little is of any use to consumers. It does reveal that most people would have to be eating very unusual diets if their intake of colours were to exceed official standards; but it also reveals that it would not be too difficult to exceed officially recommended maximum levels of intake for caramel colours. The report fails to reveal, however, how much of which colours are in which products. The Government does apparently possess this information, as it was acquired in the preparation of that report, but declines to share it with the public.

As to the safety of colourings, permitted colourings have all been extensively tested on laboratory animals. The information gained from these tests does not yet show that these synthetic colours are safe for humans. On the contrary, we know that for a (probably small but as yet) undetermined proportion of the population some colouring additives can cause unpleasant acute symptoms, such as rashes, asthma and hyperactivity. On the basis of tests that have been conducted with animals we know only how safe they are for rats, mice and hamsters. Doubts remain particularly about the chronic effects of many colours for humans.

I estimate that currently the British food and drink industry is spending about £2 million a year to buy 700 tonnes of natural colours, £5 million a year to buy a similar quantity of synthetics, and £8 million to buy 15,000 tonnes of caramel colours.

There are a few other types of organoleptic cosmetics such as acidulants and buffers, but they do not warrant any detailed treatment here. There remain the textural cosmetics, which do deserve consideration. The main group of chemicals which perform this function are the emulsifiers, and they are very important to the food industry. Emulsifiers are used to make oil and water mix together, which is otherwise not possible. Products like margarine and 'non-dairy ice-cream' would rapidly separate out into water and oil layers if they did not contain chemicals which helped to produce an emulsion. Chemicals which facilitate emulsions are the emulsifiers, but their action generally has to be supplemented with chemicals called stabilizers which prevent the emulsions from breaking down.

Altogether, fifty-seven different chemicals are permitted as either emulsifiers or stabilizers, with E-numbers in the 400s, plus lecithin (E 322). I estimate that the UK food industry is currently buying no less than 16,000 tonnes annually of these chemicals for approximately £16 million. Other textural modifiers include jelling agents, gums and thickeners.

Another significant group of food additives are the modified starches. They are important for at least three reasons: first, they are used heavily in tinned baby foods and consequently can constitute a high proportion of an infant's supply of carbohydrates; second, they are there essentially for commercial reasons only; and, third, their use in the UK is not regulated.[14]

4 Processing aids

Some chemicals such as anti-caking agents are added to mixtures of powders so that they will flow smoothly through the tubing of factory equipment. Lubricants are added to fluid mixtures to try to prevent the products from sticking to baking tins, or to other processing equipment. Other processing aids include: enzymes, clarifying agents, anti-foaming agents, solvents, release agents and neutralizing agents to reduce acidity.

These apart, there is one particular group of processing aids which deserve special comment, and these are the polyphosphates. Polyphosphates are used primarily in processed meat and fish products. They are used essentially as water-retaining agents, because they cause the flesh of animals and fish to behave like sponges. If you inject the flesh of a chicken with polyphosphates before you freeze it, you can change the cells of the tissue to make them swell up with water, and still not appear wet. Polyphosphates are used for this reason in many hams, sausages and processed meat and fish products. It was the use of polyphosphates in meat products which occasioned the wonderfully honest advertising slogan 'Why sell meat when you can sell water?'[15] I estimate that currently the UK food industry is spending £5.5 million a year to buy 10,000 tonnes of phosphates, of which about one third are polyphosphates destined for meat and fish products.

5 Nutritional additives

Nutritional additives are mainly vitamin and mineral supplements whose presence is usually conspicuously proclaimed on the packet. Something in the order of £1 million is spent annually by British food companies to purchase these chemicals. There is legislation which requires the 'fortification' of bread flour and margarine with some nutrients. As a result white bread is 'enriched', as it is euphemistically called, with some of the nutrients that are removed in the flour milling. (The scientific committee which advises the government on this matter views this as only token enrichment because the nutrient levels are not returned to the values that are present in wholemeal bread.) There is evidence to show that the practice of enrichment is beneficial to some large, poor families and to the elderly. There is also evidence suggesting that our intake of some essential vitamins and minerals has been declining since the Second World War. This has mainly resulted from an increased reliance on refined processed foods, which often have a diminished nutritional quality. One can conclude therefore that the problem is not that of reintroducing some of the goodness removed by processing, the problem is that it has been removed in the first place. This problem is compounded and made more acute by the deleterious nutritional consequences of the actions of industrial plant-breeding stations. In pursuit of heavy-cropping, machine-pickable and fertilizer-responsive varieties, commercial breeders have ignored the nutritional value of their products, which has consequently deteriorated.

The costs and benefits of additives

Of the £225 million (+/− £25 million) which I estimate the UK food-processing industry to have spent on food additives in 1985, less than 1 per cent goes to buy consumer protection, about 1 per cent goes to buy nutrients, about 10 per cent buys processing aids, while the remaining 88 per cent is spent on cosmetics.

To what extent does the use of these additives benefit the food industry and consumers? It is not possible to provide any comprehensive overall estimate, but for a handful of examples fairly precise figures are available. Consider the chemical sodium nitrite which is used as a preservative and is incorporated into all bacon and ham, as well as in a wide range of cooked meat products. I estimate that in the UK meat processors are spending just a bit less than £40,000 (at wholesale prices) to buy some 150 tonnes of sodium nitrite, which is used to produce approximately 1 million tonnes of nitrite-cured meats. The average British consumer eats about 1.5 ounces of nitrite-cured meat daily. As a result pig farmers are earning almost £1 billion, with a bit less than that being earned by the meat-processing industry.

Consider also the two most commonly used synthetic antioxidants, namely BHA and BHT. The total UK annual demand for these chemicals from the food industry is about 100 tonnes, at a combined cost of about £100,000. The use of these antioxidants sustains, amongst others, a fried-snack-foods market valued at about £660 million a year.

These examples demonstrate that the use of additives can be of enormous economic benefit to the food industry, as well as to farming. But can they benefit consumers too? I believe that there *can be* benefits to consumers from the use of additives, but only if they are demonstrably safe beyond reasonable doubt, and if consumers are fully informed about the composition of products, and their nutritional and toxicological status. In practice, however, as a leading professor of toxicology has pointed out, food additives are not incorporated in products because consumers asked for them, their presence is often concealed, it is hard for individual consumers to avoid them, and their use can exploit uninformed patterns of taste and choice, and lead to a loss of information.[16]

Enough has been said to show that the benefits to the industry from its use of additives are substantial, and substantially outweigh any benefits which their use may confer on consumers. If there are risks which arise from the use of additives, these risks are borne almost entirely by consumers, and not by manufacturing companies. To decide whether or not there are risks to consumers from additives we need to consider how additives are regulated and tested.

3 | Who Controls the Chemicals and How?

The public expects, and is entitled to expect, the government to regulate the use of food additives so as to protect us from any toxic hazard which might arise from their use. We are also entitled to expect access to sufficient information to enable us to judge how effectively the government is fulfilling its responsibilities.

If we were just to examine the rhetoric of the law we could be forgiven for thinking that everything is perfect, and that there is nothing to worry about. The law says that:

No person shall add any substance to food, abstract any constituent from food, or subject food to any process or treatment, so as (in any such case) to render food injurious to health, with intent that the food shall be sold for human consumption in that state.[1]

There is very little to complain about in that statement of principle, except that it would be desirable to revise the final clause to extend it to cover free samples as well as food for sale. Section 4 of the Act: 'Requires Ministers to have regard to the desirability of restricting so far as practical the use of substances of no nutritional value.'[2] Unfortunately the reality fails to match the rhetoric by a very wide margin.

In classical and medieval societies the adulteration of food was extensively regulated, and miscreants were often severely and conspicuously punished. The industrial revolution which started in the 1780s in the textile industry of northern England resulted in massive movements of people from the countryside into the cities, and the urban population provided the embryonic food-processing industry with its market, and with incentives and opportunities for the increasing adulteration of food products. This period was also the time when British governments became infatuated with the ideology of the free market. Adam Smith and other classical economists did not have a great deal of difficulty in persuading the urban bourgeoisie that consumers' interests would be best served by dismantling restrictions on production and trade and giving free rein to market forces.

As a result, in 1815 Parliament repealed the last remaining residue of mediaeval food regulations when they dismantled the Assize of Bread, which for centuries had controlled the composition and price of loaves. From 1815 until 1875 the food markets in Britain were entirely unregulated. New regulations were eventually reintroduced, but only with a great deal of difficulty, and after a lengthy struggle. The government was reluctant to accept the responsibility for controlling the actions of the food industry, and the food industry campaigned very effectively to discourage the government from doing anything more than the minimum necessary. Something was necessary though because there was a long series of increasingly outrageous scandals involving the fraudulent and toxic adulteration of foods, and because there was a sustained public campaign demanding regulations.

The 1875 Sale of Food and Drugs Act required local authorities to appoint public analysts, and empowered local inspectors to obtain samples for analysis. The first statutory regulation specifically concerned with food additives, as opposed to adulteration or contaminants, was introduced in 1925 prohibiting the use of preservatives in a broad range of foods. The first comprehensive legislative system for additive control came in the 1938 Food and Drugs Act. Although that empowered Ministers to prohibit the use of several different groups of additives, it still did not give them the authority to establish positive lists, or to require evidence of need and safety before an additive could be used. For this we had to wait until the 1955 Food and Drugs Act, and this remains the basis of current legislation although it has been tidied up in the 1984 Food Act.

The governments of all the industrialized and some Third World countries control the use of additives. No two countries control additives in the same way, and there are major differences in both institutions and regulations. Not all of these differences can be explained by reference to the profoundly uncertain scientific basis of regulatory policy as they also reflect the differing relative strengths of industry and consumer lobbies. In Britain, there is no effective consumer lobby on this issue, but the industry is well organized, strong and effective. As a result there are fewer restrictions on the use of additives in the UK than in other comparable countries.

The basis for the control of additives in Britain is that governments must first pass laws which give them the power to make specific regulations. Specific regulations are then introduced by governments on the basis of advice from a group of selected experts. The industrial and commercial use of food additives in Britain is controlled (in so far as it is) with regulations issued by the Ministry of Agriculture, Fisheries and Food (MAFF). The laws which are used empower, but do not oblige, Ministers to make regulations. Ministers have a great deal of discretion, for although they are advised by committees of experts they do not have to accept their advice, and sometimes that advice is rejected. Furthermore, after receiving that advice they consult with at least some of the interested parties, that is to say primarily with the food industry, and the government may then choose to modify or even abandon any proposed regulations. When Ministers have decided on some regulations these are then laid before Parliament, and almost invariably go through on the nod. The arrangement is that regulations proposed by the government to Parliament become law automatically unless they are challenged by MPs. The regulations do not have to be debated, and hardly ever are. To the best of my knowledge, only once has a proposed additive regulation been challenged in the House of Commons, and that was on 9 November 1967 when Joyce Butler MP forced a discussion on the acceptability of the artificial sweeteners known as cyclamates.

When MAFF was given responsibility for controlling the use of additives it appointed two advisory committees. Any decision as to precisely which industrial practices to permit and which to forbid is a difficult one. One of the main reasons for this is because there is a great deal of uncertainty in the science of toxicology; there are also profound conflicts of interest between consumers and the food industry. It is in the interests of industry to have as few restrictions as possible on their use of additives, but consumers want to be confident that they are fully protected and are receiving the benefit of all the scientific doubts. It is therefore important to know the extent to which industrial and consumer interests have been, and are being, represented and defended in the expert committees and ministerial meetings.

The two main committees which advised MAFF were called the Food Standards Committee (FSC) and the Food Additives and Contaminants Committee (FACC), but in 1984 they were combined into the Food Advisory Committee (FAC). The membership of these committees was chosen by the Ministry with no official reason whatsoever given as to why particular people were, or were not, chosen. This is still the case with the FAC. We are, however, told that individuals are not chosen as representative of particular interest groups. This is a fiction which appeals to the secretive state, but which does not survive even cursory scrutiny. When we look at the names and institutional affiliations of those who have served on these committees over the years we can see a familiar pattern. People are chosen if they are considered scientifically reliable, but only if they are willing to accept the constraint of current practices and policies.

In keeping with their habitual practice, ministers are well advised by their civil servants as to who could be useful, but also on who might be likely to rock the boat, and who can be relied upon to act responsibly as the government would see it. The membership of the committee includes medical scientists, senior public analysts, directors of food research stations, plus several senior executives from leading food companies, and one token consumer representative. The companies who have been directly represented on the committees have included: Marks & Spencer, Unilever, Reckitt & Coleman, Grand Metropolitan, Cadbury-Schweppes, ICI, Rowntree-Mackintosh and Smedley-HP Foods. Even if we could have counted the token consumer representative as a genuine spokesperson, consumers would be overwhelmed by industry representatives by a factor of five to one. To appreciate fully how these committees have treated the scientific uncertainties and conflicts of interest, we shall examine their decisions as well as their membership.

There are two basic ways in which additives can be regulated. The weaker system uses what is termed 'a negative list', while the stronger system uses 'a positive list'. In 1976, the FACC recommended that flavourings should be covered by a small negative list. This would have meant that twelve flavouring substances would have been banned, while all other flavouring could be used. This proposal has never been implemented. In contrast, preservatives and colours are covered by positive lists. This means that a colouring or preservative may only be added to food if it is specifically listed and permitted, otherwise it is banned. The crucial difference is that for an additive to be placed on a negative list, the government must prove that it is harmful before action can be taken. With a positive-list system it is the responsibility of industry to satisfy the government that a particular additive is needed and that it can be safely used before it is permitted. Only with a positive-list system is there any requirement to pre-test an additive before it is introduced into our foods.

It is obvious that positive lists serve the interests of consumers more effectively than negative lists. The first group of additives to be regulated in Britain with a positive list was the colours, starting in 1957, and that change alone reduced the numbers of colours in use from eighty to thirty. The UK has consistently trailed behind almost all other industrialized countries in the establishment and extension of the use of positive lists. If a food manufacturer, or an additive manufacturer, wants to introduce a new additive which falls within a regulated category, or to initiate a new use for an existing additive, then they are required to obtain permission from the Ministry. The Minister refers the issue to the advisory committees. These are two criteria which the committees are supposed to use when making their judgements and recommendations: an additive should only be permitted if there is: (1) a need for it; and (2) it can safely be used. These two criteria may seem very simple in principle, but in practice they can be extremely controversial.

Are additives necessary and if so for whom? Consumers need wholesome safe food, and if it can be shown that additives are necessary for that purpose then to that extent consumers do need some additives. Consumers do not, however, by any stretch of the imagination need cosmetic additives.

Does industry need additives? Strictly speaking they do not need additives any more than consumers do. If the use of additives were controlled far more strictly then there would be only a few losers but several winners in the industry. The main losers would be the companies who manufacture the additives, rather than the food companies themselves. British manufacturers could reap substantial rewards from improved additives standards, although this is something which few of them appreciate. Currently, because UK regulations are relatively weak, other European firms can sell in the UK market more readily than UK firms can sell in Europe. Since the British food industry is highly concentrated and well endowed with research and development facilities, it could benefit from the increased export markets which higher British standards would provide. Unfortunately few British manufacturers recognize these opportunities and most fail to appreciate what is in their own long-term interests.

The only sense in which food processors need additives is if their direct competitors are using them. In so far as the use of additives can lower production costs, the forces of commercial competition will ensure that the use of additives is contagious. When one firm starts, the others have to follow; and each firm has an incentive to be the first in their field to take advantage of any potential new cost savings which an additive may provide. To this extent, firms have at most a conditional need for additives, but that is still not a real need. Each firm only needs them if other firms are using them. There are certain kinds of products such as margarine, ice-cream, sausages, soft drinks and sweets that could not exist in their current form without additives. But in this context the question is whether we need those products made in those ways, rather than whether we need additives as such.

Industry does not really need additives, but additives remain extremely useful to industry. How does the FAC (and the FACC before it) interpret the concept of 'need'? The need for colours has always been the easiest to challenge. In 1979 the FACC produced a report on colours which at least attempted to address the issue of need. They recognized that colours are useful to industry, and that coloured foods are easier to sell than pale foods. They could not bring themselves to claim that there is a need for additives, however, and retreated merely to claim that 'colours have a part to play in the food industry', and that their use should continue to be permitted.[3]

The advice which has been given by these committees and the decisions which have been taken reveal that the criterion of need which governs these decisions is not 'needed by consumers' but rather 'useful to industry'. The current chairman of the FAC said in 1984 that he and his committee approached their work with the assumption that there are no fundamental or long-term conflicts of interest between industry and consumers; and the assembled industrialists all concurred.[4] Thus they are able to satisfy themselves that what is good for producers must, somehow, be good for consumers too.

Questions concerning the safety or toxicity of additives are of even greater importance than questions about need. Issues of safety are not left to the FAC, but are referred to committees of the Department of Health and Social Security (DHSS). The DHSS committee most directly concerned is the Committee on Toxicity (COT); COT is supplemented by the Committees on Carcinogenicity and Mutagenicity (COC and COM). These committees consist mostly of medical professionals, but also include industrial employees and consultants, although no consumer representatives. It is very hard to discover very much about any of these committees because their activities are concealed behind the Official Secrets Act. The FACC, now a part of the FAC, does produce some reports, and sometimes they include in an appendix summaries of the reports which they have received from the DHSS committees, but these documents are bland and uninformative.

The committees of MAFF and the DHSS primarily consider information provided by firms seeking permission to use additives. Understandably, it is government policy that the firms who are seeking permission for the use of an additive should pay for the testing of that chemical. As a result the UK government does not carry out any safety evaluation work of its own whatsoever. Firms either conduct the work themselves, or pay someone else to do it for them. They may turn to a commercial laboratory or the British Industrial Biological Research Association (BIBRA), the industry research association. In the USA, as in other countries, the regulatory agency the Food and Drug Administration does pay for, and conduct, some safety evaluation work itself. But the British government neither funds, conducts nor commissions any safety evaluation work on additives.

This does not quite mean that industry provides all the information which the committees review and assess, because they also review the data available from other national regulatory agencies, as well as from the Scientific Committee for Food (SCF) of the EEC, and from the international Joint Expert Committee on Food Additives (JECFA) established by the UN Food and Agriculture Organization and the World Health Organization. But these organizations too obtain their data essentially from industry.

A small fraction of this safety and toxicology data is available in the public domain, but most of it remains secret. There is an extensive published literature on food-additive toxicology, but the indications are that no more than 10 per cent of the gross volume of information presented to governments is ever accessible to the public. The worst case of gratuitous concealment concerns the artificial sweetener acesulfame-K. This was first introduced into the British market in the autumn of 1983 despite the fact that no toxicological information on it whatsoever had been published in Britain. Food-industry scientists will complain that it is difficult or impossible to persuade journal editors to publish scientific papers reporting the results of tests in which no interesting effects were found. While this may be true, there is nothing (short of intense political pressure) to prevent the government from requiring that all data submitted in evidence to the regulatory authorities should be available for public access in libraries.

When it comes to the question of how the FAC and the DHSS committees operate, we really know very little because so little is revealed. The official account says that the committees seek to reach a consensus amongst the members. What this means is that conflicts of interest are being dealt with by minimizing their significance. The assumption is that it is possible for a committee to determine what is in 'the public interest', as though there was one and only one shared communal interest, rather than two or more conflicting points of view. Since the dominant interest group influencing the deliberations of the committees is industrial, they will readily assume that what is in the interests of industry is in 'the public interest'. In practice, therefore, insufficient consideration is given to consumer interests, and industrial interests massively predominate. In public the food industry proclaims its determination to conduct itself strictly in accordance with all laws and regulations – but, behind closed doors, it is industry which plays a dominant role in setting the regulations which it then follows.

I once heard a senior food-industry representative complain that the industry was burdened by excessive and unreasonable regulations and that 'we' (the industry) should determine the regulations under which it operates. I have no doubt that the industry does not entirely get its own way, and that they are pressed into accepting more restrictions than they ideally would choose, but it remains the case that, judged by the standards of consumer protection, and by comparison with other countries, British additive regulations are unacceptably weak, and biased in favour of industry.

Political scientists (as they like to call themselves) have introduced the expression 'regulatory capture' to describe a situation in which the industry being regulated has gained control of the agency which is established to regulate it. The extent to which the UK regulatory agencies have been captured by industrial interests can only be determined once we have reviewed the strengths and weaknesses of toxicology, but we already know enough to be able to see that the institutional arrangements readily lend themselves to this eventuality.

At the end of their deliberations over the possible use of an additive the advisory committee publishes a report in which it outlines its discussions and lists its recommendations. There are four types of recommendations which the FAC can provide. First it may recommend to the Minister that some particular additive be permitted, usually for some specified purpose, and occasionally subject to quantitative restriction but more generally subject to what is termed 'good manufacturing practice'. A second kind of recommendation would be that some additive should be temporarily permitted, but that permission should be reviewed in the light of further evidence by a specified date. Third, the committee may recommend that permission for the use of an additive should be withheld pending the availability of more evidence; and, finally, a recommendation that an additive should not be permitted may be given. Following the publication of a report the Ministry invites comments from interested parties, before proceeding to propose regulations. Only after receiving those representations will Ministers announce draft regulations. Before these regulations are put before Parliament a further round of consultation occurs. The documentary record shows that firms readily avail themselves of the opportunity to influence these ministerial decisions, while consumers hardly ever do.

It may come as a surprise to most people to discover that only a minority of permitted additives are subject to quantitative restrictions. Apart from a few special cases arising mainly from compositional standards in products such as bread, only the preservatives and antioxidants are covered by regulations which specify quantitative restrictions. Maximum levels of usage are specified for these additives because they are most open to abuse. High levels of these additives could be used to conceal the age and relative decay of foods. As for all the rest, manufacturers are, as stated above, only 'subject' to good manufacturing practice (gmp).

When I asked the MAFF office responsible for controlling additives who decides what constitutes 'gmp', an official told me that it is up to the manufacturers to decide what they judge to be good practice. This situation makes life extremely difficult for enforcement officers such as those responsible for environmental health and consumer protection. Enforcement officers have a very important job, made more difficult by the fact that cuts in local authority spending have reduced staffing levels at precisely the same time as their potential work-load is increasing. This is even more deplorable when the technical demands of the job are becoming more exacting, so that not even funding maintained at a constant level would enable public analysts to keep up with technical changes in manufacturing and chemical analysis.

If an additive is covered by quantitative restrictions governing the levels at which it can be used, it can be relatively straightforward for enforcement officers to determine whether or not those levels are being exceeded. But where the regulations specify only 'gmp', then the enforcement officers are not in a position to judge whether any particular manufacturing practice is good or not. Life is made even harder by the fact that, under British law, enforcement is at the point of sale and not at the point of production. This means that often enforcement officers do not know what the actual practices of manufacturers are, let alone whether or not they are 'good'.

Food safety is the primary and dominant consideration in food-additive regulation, but it is not all that has to be considered. The other major concern is the labelling of manufactured food products. The requirements for additive labelling under British law are relatively lax by comparison both with other countries and with what consumers might require and desire. The philosophy underlying British policy appears to be that as the government is looking after the safety of food, consumers do not need to concern themselves about composition, and therefore do not require, nor are they entitled to, complete information. The government assumes for itself the responsibility of deciding how much information consumers want and need.

In 1984 I heard a senior and influential British academic nutritionist discussing the public's attitude to food labelling. It was his considered opinion that consumers know and understand so little about food and health, that they are not interested and they do not want what they cannot understand, and so labelling should not give more than a minimum of information. He did not seem to appreciate that the extent of ignorance among the British public is itself a consequence of industrial secretiveness, compounded by the condescending attitude which he was himself exhibiting. While many junior members of the nutrition profession have made commendable efforts to inform the public, the leadership of the profession, in close alliance with the food industry, has preferred to maintain their exclusive grip on food science.

In 1983 a home economist working for a major retail chain revealed on Radio 4's 'Food Programme' that her company was convinced that if they listed all the ingredients on some of their products then it would become impossible to sell them. That admission was refreshingly frank, and it was even more refreshing to hear in 1985 that the same firm had reversed its policy and was taking the initiative by improving their labelling, and reducing their reliance on additives. A succession of surveys has revealed that although consumers do not have much knowledge about additives, they are very interested and do want to know a great deal more.[5] Labelling regulations have, therefore, an important part to play.

In Britain it was not until 1974 that food-product labels had to carry any compositional information whatsoever, but since then partial labelling of some products has been required. Before labelling regulations were introduced in Britain the food industry put up sterling and well-organized resistance to proposals for improved labelling. Throughout the intervening period large sections of the industry have fought to limit the extent and scope of the labelling regulations. A small minority of firms have taken the lead and improved their labelling, and may have benefited as a result, but most of the industry remains recalcitrant.

There are many products, such as bread, which is sold unwrapped, and sweets, which were sold with small wrappers only, that do not have to be labelled at all. The number of exclusions from the labelling regulations are so extensive that the document which comprehensively lists all the regulations is vast. The regulations are so complicated and detailed that it is quite unrealistic to expect consumers to be able to tell when and which additives are in which products yet not on a label. While the retail trade is covered by a complex mass of rules, the catering trade is entirely excused them. As a result it is impossible to know what we are eating in restaurants and canteens. The processed-food products which are specifically designed, manufactured and sold to the catering trade are particularly liable to contain additives. The fast-food sector is especially reliant on additives, and even the so-called 'gourmet' end of the market is not immune from this problem either.

From 1974 until 1983, for those products which did have to be labelled, manufacturers were only required to use category terms like 'colouring' or 'preservative', without having to name particular ingredients. As recently as 1981 the British government, at the behest of the food industry, had set its face against comprehensive improvements in labelling regulations. Eventually in 1983 improvements in labelling came about as a result of initiatives taken by the EEC, and in the face of strenuous opposition from industry. Since 1983 the law has required products to list the name or the EEC E-number of many of the additives used. The value of this requirement is limited, of course, by the fact that many types of products are excluded from these rules, and because flavourings are not subject to the requirement.

A senior industrial scientist has discreetly pointed out that the British food industry only agreed to accept and co-operate with the E numbering system when they were threatened with a ban on the colouring additive tartrazine, along with other chemicals known to provoke intolerant symptoms in sensitive people. The industry accepted the argument that if customers could not identify products containing these additives, then they would be unable to avoid them. Industry was faced with a choice between a ban or better labelling. They chose, what was to them, the lesser of two evils.

There will be a major step forward on 1 July 1986 when all products (if they require labels) have to list each additive by its E number or its category name, and in a sequence which reflects its proportion by weight. However, that is not quite the breakthrough it sounds, and for several reasons. First, too many products do not have to be labelled at all and, second, flavourings do not have to be individually identified. Third, listing them in a sequence which reflects the quantities involved is not as good as a precise statement of the amounts of each chemical present. As Caroline Walker, one of Britain's leading nutritionists, has frequently pointed out, there are no earthly reasons why we should be entitled to more information about the composition of our socks than of our sausages. Textile labels tell us the percentage by weight of the ingredients. Food labels should be no worse than those.

The first occasion when the UK government provided even a partial list of additives in use was in January 1983 when they first published a list of E numbers in a leaflet entitled *Look At The Label*.[6] The first edition was, however, incomplete since it did not include all those permitted additives which do not have E numbers. It was extremely difficult, but just possible, to obtain a list of permitted additives to which no number had been assigned. More recent editions of *Look At The Label* have combined both lists, but they are still rather uninformative. An alternative pamphlet entitled *Look Again At The Label* identifies regulated additives by name and number and divides them into three columns headed 'beware' 'suspect' and 'safe'.[7] This is a very useful document which deserves a wider distribution.

One reason why the more enlightened manufacturers and retailers are starting to think about the desirability of producing either additive-free products, or products containing fewer additives, is because when the E numbers and names started to appear on the labels some consumers understandably but mistakenly assumed that the additives had just been introduced into the products. One consequence was that marketing personnel went to the production staff and told them that they should try to reduce the use of additives because consumers did not want them. When the food industry started to complain about what to them were irrational consumers, they were told that it was their own fault for having been so secretive in the first place and for so long.

It would be unrealistic to list the full chemical name of all additives for all products on the labels of food packaging, because some chemical names are very long and there is just not enough space. Consequently having an E-number system is probably quite sensible, although it is altogether too difficult to obtain a list of E numbers, and too many additives are permitted without having an E number. Great improvements still remain to be made in UK labelling regulations, which need to be more extensive and more comprehensive.

One further requirement under food regulations is that illustrations on packages should not mislead consumers. It is illegal, for example, to sell an orange-flavour drink with an illustration of an orange on the packet unless there is some real orange in the product. As a result, the company which manufactures the 'table-top' sweetener called Canderel, which contains aspartame, were obliged to change their packaging when it was initially launched with an illustration of an apple and an orange.

There is in Britain currently an official discussion about the need to improve compositional and nutritional labelling but the debate is focused almost exclusively on fat labelling, to the neglect of additives and of sugar, salt and fibre. It is hard to see why it should not be illegal to introduce any chemical into our food supply without a full declaration to the government and to consumers. Food-additive labelling in several other countries is far superior to that in the UK and British consumers are less well informed than they both could and deserve to be.

To whom are the UK regulatory authorities answerable, and how do their operations compare with those of other countries? Additives in the USA are regulated by the Food and Drug Administration, and the actions of the FDA are subject to scrutiny by Congress and the courts. The FDA can be forced by public-interest groups to conduct public hearings on particular regulatory decisions, and Congress is entitled to review and cross-examine the authority and to call any other witnesses it chooses. Furthermore, individuals and pressure groups can challenge the decisions of the FDA in the American courts. All of these procedures have been used from time to time, and this possibility tends to keep the FDA on its toes. No similar provisions exist in the UK.

Strictly speaking, the Minister of Agriculture, Fisheries, and Food is answerable to Parliament for all departmental decisions, but with the exception of the unique case of cyclamates, Parliament has not reviewed or questioned the actions, judgements or operations of the FACC or the FAC. In principle, the House of Commons Select Committee on Agriculture could establish such a review, and it would be a very good idea if they would do so, but they show no sign of such action. Finally, there is no provision under British law enabling consumers to challenge any action of the regulators. It has recently been decided, however, that those who suffered harmful effects from the drug Opren, and their surviving relatives, should sue the DHSS and the Committee on the Safety of Medicines. If that action comes to court, and if it is successful, then it will set a precedent for a whole set of challenges to many UK regulatory authorities.

It will only become possible to make detailed judgements on the specific regulations which are in force after a discussion of the testing of additives, but some readers might be tempted to suppose that regulations can be no better than the toxicological knowledge on which they are based. That would, however, be a mistaken oversimplification. There can be better or worse ways of coping with uncertainty and with inadequate information. This is a question, therefore, to which we shall return.

4 | How Are Additives Tested for Safety?

The science of toxicology: our saviour or a shambles?

When we want to try to establish whether or not some chemical is safe for human consumption it is to the science of toxicology that we turn. Toxicology is the study of the ways in which chemicals may be safe, poisonous or therapeutic. All the relevant parties, namely governments, industry and consumers turn to toxicology to try to discover what is safe and what is not. Life would be so much simpler if toxicology were an uncontested science, but we are not so fortunate. It is not just that toxicology is a contested science but, worse, it is probably the most contested of all the sciences. It might, in principle and in an ideal world, be possible for toxicology to stand outside the conflicts of interest between industry and consumers, but in practice toxicology is a highly politicized discipline. Toxicology is neither a pure science, nor an innocent science, but rather it manages simultaneously to be an instrument, an agent and a victim of the conflicts in the field of regulatory policy.

Ideally we would want toxicology to be a precise and definitive science because if we knew exactly what was and what was not safe it would be far more straightforward to decide which industrial practices to permit and which to forbid. Unfortunately, toxicology is an extremely imprecise discipline and it is very rare for it to provide accurate and unambiguous answers to practical problems. The world in which we live, moreover, is a long way from being ideal, and in this real world there are circumstances in which people are eager to identify and exploit the uncertainties of toxicology, as and when it serves their interests to do so.[1] This chapter will identify and assess many of the uncertainties in toxicology, not to exploit them but to explain the ways in which they can be, and are being, exploited.

When we try to assess the safety or toxicity of an additive there are just three main sources of information to which we can turn. These three areas of toxicology are firstly human epidemiology, secondly what are called short-term *in vitro* tests, and thirdly tests with live animals. Each sub-discipline has its own particular contribution to make, and each has its own limitations. As and when we have obtained data from these three sources we may be in a position to evaluate the safety or toxicity of a chemical for human use. Inevitably, it is difficult for these evaluations to be less problematic than the data upon which they are based.

Whatever limitations there may be on the science of toxicology, it is nevertheless expensive to test a food additive. It costs approximately half a million pounds to generate sufficient data to enable the government to reach a regulatory decision. Industry is only willing to invest that much if they believe that they have a chemical for which there is a substantial, reliable and profitable market.

Human epidemiology

Human epidemiology is the study of the variations in the patterns of illness and death which occur between different groups in the population. Epidemiologists seek both to chart those variations, and to correlate them with environmental factors which may account for the observed patterns. In one crucial respect epidemiology is the most fundamental of the three components of toxicology. This is because epidemiology studies human beings, while the others scrutinize non-human organisms. Veterinary science apart, the only point of engaging in non-human toxicology is to help us to know more about the health and safety of humans. Non-human toxicology may have a part to play by providing us with what are called 'model systems': non-human systems which are intended to model human biology. But the adequacy of any particular model can only be judged by the extent to which it supplements our knowledge of human biology. Some people say that we have to use non-human model systems such as laboratory animals and bacteria because it would be immoral to conduct experiments on human beings. Others more cynically say that we are conducting experiments on human beings all the time, and human epidemiology is merely the business of monitoring the results of those experiments. Our knowledge of the toxic effects of chemicals on humans derives essentially from the science of epidemiology. This is why epidemiology is the most important component of toxicology, because epidemiology provides the standards by which all other contributions must be judged.

Epidemiology has been, and remains, an important science, despite its very severe limitations. For example, it is epidemiologists who have shown that smoking leads to lung cancer and heart disease, that nuclear radiation causes leukaemia, that salmonella and botulism are acutely poisonous, and that industrial chemicals such as asbestos, coal dust and vinyl chloride are killers.

The methods of epidemiology can be very useful when we want to identify the causes of an outbreak of acute illness, but they are far less effective when we need to identify the specific causes of chronic problems. The methods of epidemiology are sufficiently sensitive to enable us to identify the causes of most cases of acute food poisoning. As a result, cases of acute food poisoning must now be counted as arising from failures to apply the knowledge which we do now possess. Unfortunately, when it comes to the issue of assessing the chronic toxicity of food additives human epidemiology has very little to contribute.

What human dietary epidemiology can accomplish for example is to show that most British people, as well as other Europeans and Americans, would be healthier if they were to eat less fat and less refined carbohydrate, cut down on their salt intake, and have more fibre in their diets. Epidemiology can be useful in studying occupational health problems. Where it is possible accurately to distinguish a sub-group of the population that is exposed to high concentrations of particular chemicals over long periods, then toxicity is relatively easier to identify. However, when it comes to estimating the long-term consequences from the use of food additives, epidemiology is almost entirely useless. With some 3850 or more additives used in thousands of products in millions of combinations in small quantities (though in some cases using up to thirty different additives in a single product), it is impossible for epidemiology to identify any long-term or chronic effects of the use of particular food additives.

Even where epidemiology is informative it is only able to identify a chemical as a poison after extensive harm has been done. For this reason, epidemiology can make no contribution to the pre-testing of food additives. If a food additive which is already in use were acutely toxic to more than 10–15 per cent of the population then it is likely that we would already have established that fact. We can therefore probably be confident that no food additives are being used which make many of us ill for much of the time. If, however, a food additive were acutely or mildly toxic to less than 10 per cent of the population, then we may well not yet have discovered that fact, and so may still be using such chemicals, and causing some illness and distress. Where there are chronic hazards from food additives, epidemiology cannot detect them, and so it cannot pronounce on either safety or toxicity.

This is important because sometimes the results of animal tests show that an additive which is already in use is chronically toxic to laboratory animals. When confronted by that evidence it is then always possible for an industrial spokesman to say: 'But there is no epidemiological evidence that it is causing any harm.' Although this is almost invariably true, this is solely a reflection of the fact that if there were chronic hazards to humans, epidemiology would be incapable of detecting them. If there were any epidemiological evidence of the toxicity of an additive then this would be conclusive grounds for banning it, albeit too late for those who had already suffered, but the *lack* of epidemiological evidence of chronic toxicity does not even start to be evidence of safety in use.

Short-term mutagenicity tests

The toxicology of non-human model systems is divided into two parts: there are tests with whole live animals, and there is a group of tests which use tissue and bacterial-culture techniques, and are conducted in glass dishes. Scientists call these '*in vitro*' tests, although sometimes they are loosely referred to as short-term tests.

That description is slightly misleading because the important distinction is between those which are carried out in glass dishes, and those conducted on live animals or '*in vivo*', and confusion may arise because there are some short-term tests conducted on animals. The tissue and bacterial-culture short-term tests are relatively novel techniques. Toxicologists have studied the effects of feeding chemicals to animals for over a hundred years, but the first of these short-term tests was only invented as recently as the early 1970s. There were two good reasons for scientists to search for tests which could provide alternatives to feeding studies with live animals. First, animal studies are very time-consuming and expensive and, second, while they may tell us a great deal about animals they tell us little about human beings.

The first breakthrough in the development of these short-term tests came from an American team led by Bruce Ames who were studying mutations in bacteria. Mutations are changes in genetic material, and are important for several reasons. On an evolutionary time-scale some mutations may be beneficial, but more urgently we want to know whether or not industrial chemicals are causing genetic damage to us and to our children. Chemicals which cause mutations are called mutagens, and if we ingest them then we run at least two major risks. We may have damaged babies, and we may ourselves succumb to cancer. The causes of cancer are very poorly understood, but we know that there are quite a few different kinds of mechanisms by which cancer may be induced. One clear route, however, is by causing mutagenic damage to human cells. Therefore, if a chemical is a mutagen, then it may well also be a carcinogen, a chemical which causes cancer. The short term *in vitro* tests are procedures to test for mutagenicity, and hence for the carcinogenicity of the test substance.

The breakthrough came when it was discovered that some chemicals which are known to be strongly carcinogenic to both humans and laboratory animals, such as benzpyrene and nitrosamines, also cause mutations in bacteria. It was relatively easy, quick and cheap, to detect the effect under a microscope. The optimists hoped that mutagenicity tests would rapidly eliminate the deep uncertainties of cancer toxicology. Unfortunately that has not happened, although the short-term tests do provide some useful information, as well as some evidence which contradicts some of that from animal tests. The net effect of the introduction of these short-term tests has been to reduce uncertainty in some areas, but to increase it in many others. As regards food additives, the major contribution from short-term tests has been the detection of a potentially dangerous preservative which was then being used in Japan but which had not previously been shown to be carcinogenic in animal tests.

The popularity and attractiveness of short-term tests is primarily a function of their relative cheapness and speed when compared with animal tests. They do not, however, provide straightforward answers to some of the practical problems which concern us. One difficulty is that there are now many different mutagenicity tests, and they do not all give the same results. Chemicals which are demonstrably mutagenic in one system may not show such activity in a slightly different system. This forces us to decide whether to trust the results of some tests in preference to those of others, or to try to use them all collectively and find a consistent interpretation of their results. It is extremely difficult to judge whether some, any, or all of them are relevant to specific questions of human health and safety.

There are just two basic sets of standards against which the relevance and validity of these short-term tests could be, individually or collectively, judged. We could compare them with the results of human epidemiology, and with the results of animal tests. Unfortunately there is very little relevant epidemiological data against which to compare mutagenicity tests. Mutagenic and carcinogenic poisoning are chronic problems about which human epidemiology has very little to say. Apart from tobacco smoke and nuclear radiation, we only know of some eighteen groups of chemicals which are proven to be human carcinogens from epidemiological studies. These chemicals have all been just about removed from the human environment. As a result, with the exception of smoking-related cancers, we just do not know what causes 99 per cent of the rest of human cancers.

Many chemicals (somewhere between 700 and 800) have been shown to be mutagenic to bacteria or carcinogenic to animals but we do not yet have a way of knowing which of them are toxic to humans. The debate about the usefulness of short-term tests has therefore not focused on the important question of the correlation between bacterial mutagenicity and human carcinogenicity, but on the simpler but less exciting issue of the extent of the correlation between the results of the short-term mutagenicity tests and those of the long-term animal-feeding studies. As a result, the participants in the debate about the value of mutagenicity tests too often assume that if these short-term tests correlate well with animal-test data then they are useful and relevant to human beings. Unfortunately this ignores the fundamental question about the extent to which animal tests are themselves relevant to humans.

There is a heated debate about the validity and relevance of these short-term tests, or rather the extent of the correlation between bacterial mutagenicity tests and animal carcinogenicity tests. What we find is that more than half of the chemicals known to cause cancer to animals cause at least some bacterial systems to exhibit mutations, but we also know that at least some chemicals which cause cancer to humans and/or animals do not exhibit bacterial mutagenicity. The short-term tests can only identify carcinogens which cause cancer by damaging genetic material. There are mechanisms of carcinogenesis other than mutagenic ones.

What happens in a cancer is that cells act in quite unusual ways. The usual and healthy pattern is for cells to divide at a steady rate, and then to differentiate. What this means is that newly created cells are very unspecific, in that they could for example become heart muscle, skin, kidney or bone cells. The process of becoming a specific type of cell is called cell differentiation. In a cancer the cells divide at an exceptionally rapid rate and then fail to differentiate. We have a very incomplete understanding of how and why cancers occur, but we do know that there are at least two quite different causal routes. Damage to the genetic material contained in the nucleus of the cell may cause cancer, although cancers may arise in cells with entirely normal genes, and cells can become cancerous as a result of interactions between cells, as much as from faulty nuclear material.

This is important for food-additive toxicology because it means that the most that mutagenicity tests could accomplish would be to detect a sub-set of the chemicals which cause cancer, but they could not detect any non-mutagenic carcinogens. Furthermore, mutagenicity tests are so specific that they are irrelevant to all other toxic but non-carcinogenic hazards. As a result, all we can conclude from the results of *in vitro* mutagenicity tests is that if a chemical is mutagenic to bacteria then we should regard it as potentially toxic to humans, but if no effect is found, we cannot conclude either that it is safe for humans, or even that it cannot contribute to causing cancer. These short-term tests are, therefore at best, a screen for carcinogenesis. If a chemical fails these tests, individually or collectively, we have *prima facie* evidence that it constitutes a hazard to humans, and that it should therefore not be marketed. The failure to be shown to be a *prima facie* risk by these tests is, on the other hand, not yet sufficient evidence of safety in use for humans.

Animal tests

Given the scope and limitations of human epidemiology and the mutagenicity tests, those who regulate the use of additives rely on the results of animal tests, and particularly chronic-toxicity tests, to decide which industrial practices they should prohibit and permit. It is not hard to see why this should be so. Toxicology is capable of detecting acute hazards from chemicals if those hazards are either severe or if they will affect more than about 15 per cent of the population. The question which remains therefore is whether or not industrial chemicals are chronic toxicants. If a chemical is permitted as an additive then we are liable to consume it over a long period, and therefore we need to be certain that it will be safe in long-term use, that is to say, we have to be certain that it is not chronically toxic.

Testing chemicals on animals is a complicated and expensive business. To conduct what scientists like to call a thorough 'bioassay' it is necessary to complete at least seven different major groups of tests, ranging from acute toxicity through to tests for chronic toxicity, including carcinogenicity and reproductive toxicology.

Each of the major tests involves numerous sub-tests. The term 'assay' applies essentially to the precise determination of the purity of precious metals. It is used in the context of toxicology to give an entirely spurious impression of precision and reliability to a field from which those desirable characteristics are almost entirely absent.

Cancer toxicology is a crucial component of these tests. In many respects it is the most highly developed, and the most controversial, part of animal tests. It may well be that carcinogenic hazards are not the most serious chronic problems which additives may pose, but they are the ones which receive most attention. This is partly because it is relatively straightforward to detect cancerous cells in the tissue of laboratory animals, and partly because cancer hazards are understandably a focus of social concern and so cancer research is relatively well funded by comparison with the rest of toxicology.

The standard procedure for assessing whether or not a chemical can cause cancer involves feeding measured doses of the test chemical to animals for their entire lifetime, or even feeding the chemical to a group of animals over two generations. It has been only in the last few years that a standard procedure has come to be agreed and established amongst professional toxicologists. Until the late 1970s there were just no standards in carcinogenicity testing. We may eventually have to conclude that the current procedures need to be reformed and improved, but at least in the meantime and for the most part we now know what can properly count as a standard carcinogenicity test.

Standard cancer-test procedures involve feeding the chemical to groups of fifty animals per sex and per dose level. It involves firstly conducting acute toxicity tests with similar animals to identify what is called 'the maximum tolerated dose', and then feeding that level, and two lower dose levels, to groups of animals, as well as keeping fifty males and females on an otherwise identical regime as a control. Usually some animals in each group die during the course of the experiment, while others survive until what is euphemistically termed the 'sacrifice' at the end of the experiment. During the lives of the animals their health and experience are carefully monitored and recorded. When animals die, either spontaneously or at the hands of the scientists (or perhaps more commonly, at the hands of the laboratory technicians), the corpses are subjected to detailed pathological scrutiny. Numerous tests are necessary on many parts of the animals to try to detect any adverse effects which the test chemical may have caused. To identify the effects of the chemicals it is necessary to compare the animals fed the test substance with those in the control groups, and if effects are detected it is often important to establish how they vary with dose levels.

Vital though it may be to test a chemical in bacteria and animals, this does not by itself tell us whether or not that chemical is safe or toxic to humans. When we have tested an additive we then have to evaluate that chemical. To make that evaluation involves reviewing all the evidence from epidemiology, mutagenicity tests and animal studies to see what the collective results are likely to mean for human consumers. The animal tests provide us with some quantitative data; they enable us to know with a moderate degree of accuracy how certain effects on the animals vary with the dose, under laboratory conditions.

To provide an evaluation of the toxicological significance to humans of the ingestion of a chemical is an extremely complicated business. Essentially the problem is one of extrapolation. Chemicals are tested on a small and relatively homogeneous group of healthy animals, and often at high doses, and over the relatively short lifetimes of the animals. (Mice live about eighteen months and rats a bit over twenty-four months on average.) We are trying to extrapolate from the data given by these tests to guess at the likely toxicity to a very large and heterogeneous population of an entirely different species. The human population is not remotely homogeneous. While we share a great deal, there are also great inequalities. Some people are far more vulnerable to illness and the effects of their diet than others. People differ greatly as to their genetic inheritance and environmental experience, diets and death rates, habits and habitats, practices and preferences. The population of animals on which the chemicals are tested is not just small but also narrowly uniform, and not at all like the natural populations of rats and mice, let alone like the human population. Additives are tested one at a time on relatively healthy animals. People consume additives, along with foods and other chemicals in complex cocktails, and over long periods.

It is vital, therefore, that we try to determine just how much we can learn about the effects of additives on humans from studying laboratory models. There is no consensus on this issue. This lack of consensus is not entirely surprising given the uncertainties and complexities of toxicological testing and extrapolations. When we seek to make judgements about the safety and hazards for humans from the results of animal tests we have to accomplish a long sequence of problematic extrapolations. We start with a small sample of animals, and then seek to draw conclusions for the larger population of similar animals. Sometimes the test involves using high doses of test substance over the relatively short period of the animals' lives. We therefore seek to extrapolate from high doses to low doses, and from short lifetimes to far longer lifetimes. The most problematic extrapolation is from the animal species to human beings, but in making that jump we also have to take into account the fact that human societies are very far from homogeneous. There are numerous groups in our society who may be particularly vulnerable to toxic risks. They may be very young or very old, poorly fed, unhealthy or even genetically vulnerable. Each of these factors complicates the extrapolations which we seek to make from animal test groups to human consumers, and as a result it is frequently extremely difficult to evaluate a chemical on the basis of the toxicological data.

In this context, it is not possible or even desirable to identify all the difficulties in each of the extrapolatory steps. It should be sufficient however to describe just some of the more important difficulties so that an overall judgement can be made about the value and limitations of additive toxicology.

The sensitivity of the animal tests depends on several variables. By sensitivity I mean both how effectively the sample of animals typifies the general population of animals, and how effectively the sample detects toxic hazards to that species. The more animals in each dose group used, and the more dosing levels used, the more accurate and sensitive the test can be. The number of animals used in each test group, and the number of dose levels used, is limited primarily by economic rather than scientific considerations. The size of the groups which are standardly used are such that the best which we can expect is that they can detect an effect only if it afflicts more than about 25 per cent of the animals. It would technically be a very straightforward matter to raise the sensitivity of the tests by using more animals in each group, and testing at more dose levels, but then the costs would rise proportionately. They might rise to such an extent that it ceased to be economic to develop new additives for the limited market to which they are sold.

Another important variable is what is called the background incidence of disease in the control group of animals. If the animals in the control group are always perfectly healthy then we can readily conclude that toxic effects which are found in the dosed animals are a result of the ingestion of the test substance. If, however, there is a lot of chronic illness in the control group, then it is hard to know whether the dosed animals are ill because of the dose, or for entirely coincidental non-dose-related reasons.

In practice, the sensitivity of chronic toxicity, particularly carcinogenicity, tests is very low; one of the main reasons for this is because the background cancer rates amongst experimental animals are very high. It is standard practice to provide caged laboratory animals with what is called *ad libitum* feeding. This means that food, of a known quality, composition and quantity, is freely available, and the animals can eat as much or as little as they choose. The amounts of food and test substance which they ingest can be estimated by determining the weight of food which remains from a known weight of food which is provided to the animals. In former years laboratory animals used to suffer from parasites and other pathogens and as a result they did not tend to become overweight even though they ate without restriction. Modern animal husbandry and laboratory practice has improved to such an extent that the animals are generally free of pathogens. One consequence of this is that a regime of unlimited feeding leads the caged animals to become severely overweight, relative to their natural state. Being overweight and having no opportunities for exercise can have a disturbing effect on the animals' biology, particularly their hormonal systems, and as a result they succumb readily to cancer. Relatively uncommon twenty years ago, there are now high background cancer rates in laboratory animals. Consequently it is extremely hard to detect any toxicological signal against the background carcinogenic noise.

Some toxicologists respond by claiming that the result is that we are liable to overestimate the carcinogenicity or chronic toxicity of a chemical. It is, however, just as likely that we shall underestimate the hazards. There may be a bias in the tests, but we do not yet know on which side – over- or underestimation – that bias is likely to weigh. Other toxicologists argue that background cancer rates in laboratory animals are now so high and so variable that it is almost impossible to detect anything other than highly potent carcinogens. The weak carcinogens are likely to pass undetected. Therefore if a chemical does not fail an animal carcinogenicity test we cannot conclude that it is safe, just that if it is carcinogenic, it produces tumours in less than about 25 per cent of the animals. That is, however, an extremely poor basis for regulatory decisions.

If regulatory policy is to be reliably based we would need to have a set of consistent and justified rules by which to extrapolate to humans from animal and bacterial tests. The first question therefore is: how are these extrapolations being conducted? Then we can consider the question of how justified and reliable they are.

The acceptable daily intake – extrapolation as linear scaling

In the early 1960s most governments recognized that it would benefit them all if they were to pool their expertise on the issues of food regulation. The assignment was passed to the World Health Organization and the UN Food and Agriculture Organization and they established an intergovernmental body called the Codex Alimentarius Commission, often known just as Codex. Codex subsequently established the Joint Expert Committee on Food Additives (JECFA).

It was in the context of JECFA meetings that the first standard procedure was developed for extrapolating from long-term animal tests to human regulatory judgements. A French scientist called René Truhaut was responsible for introducing and defining the concept of the 'acceptable daily intake' (or the ADI) for an additive. This figure is intended to establish a maximum average daily permitted intake of an additive by a human being. It is a specific figure which is expressed in terms of the number of milligrammes of the chemical which may safely be consumed by a human, for each kilogramme of the consumer's body weight. This figure is obtained by first identifying in animal tests a dosing level at which no adverse effects are observed, and this is called the 'no effect level' (or NEL). The ADI is directly obtained from the NEL simply by dividing it by what is usually termed a 'safety factor' (or SF). While it may be polite to call them safety factors it would be no less accurate to call them, as some do, 'uncertainty factors' or, more prosaically, 'fudge factors'.

Several major regulatory authorities are engaged in specifying and replying on ADIs. They include the US FDA, the Scientific Committee for Food of the EEC, and JECFA. They are collectively assuming that laboratory animals are sufficiently similar to humans for extrapolation to be merely a problem of scaling, and that there are no fundamental qualitative differences between rodents and humans. The crude scaling device is the safety factor which is introduced between the NEL and the ADI. It purports to take into account two considerations. First, humans may be significantly more sensitive to a chemical than are animals and, second, the human population is itself diverse, since some people are particularly sensitive even by human standards. It is conventional to say that a tenfold safety margin will be sufficient to deal with each of these factors. As a result, the standard safety factor is 10×10, namely 100. This means that it is assumed that if humans consume less than, or up to, one hundredth of the daily dose which failed to provoke detectable toxic effects in animals, then all humans should be adequately protected, and that would be an acceptable daily intake at a safe level.

All commentators admit that there is no scientific justification whatsoever for choosing this safety figure of 100, and it is generally recognized that it is chosen for political rather than scientific reasons.

The problem is not that there is some other figure which should preferably be used, but rather the illusion is that the extrapolation is simply a matter of linear scaling. The problems which confront the extrapolation from animals to humans are far more profound than can be accommodated with the simple device of the ADI.

There is, in practice, a great deal of disagreement about the value and significance of safety factors and ADIs. At least one senior industrial toxicologist is prepared to admit that the arguments about safety factors amount to an admission that we lack the information which is essential for risk assessment. In practice it has not been possible for ADIs to be specified for many additives. For many of them, there is just not sufficient information about animal toxicity to enable a judgement to be made in accordance with the standard procedures. Even where JECFA is prepared to specify ADIs, as it is for many colour additives, we find that only eleven out of twenty-five have an SF of 100, while twelve have an SF of 200, one of 600 and one of 2500. What this means is that even by the standard criteria, most colours are either insufficiently well studied or are demonstrably less safe than officials would have hoped.

The official British position is that the ADI figures developed by JECFA, and also by the Scientific Committee for Food (SCF) of the EEC, are noted but regarded with some scepticism. British regulatory practice does not rely on the concept of the ADI, despite some pressure from the Europeans that we should come into line with Continental practices. It is not that British regulators are more cautious than their Continental counterparts, it is just that our government is reluctant to provide any specific figures or to regulate additives quantitatively. This is important because if maximum permitted levels of usage are not specified it is impossible for enforcement officers to know which manufacturing practices are good and which bad.

Even if quantitative ADIs can be reliably specified they can accomplish little by themselves. Even if we have an estimate of the total quantity of some chemical which can safely be consumed each day of our lives, this still does not establish a safe level of use for particular products. If two colouring additives have the same ADI, but one is used in 100 times as many products as the other, then the amount of the first colour that can be used in each product should be no more than one hundredth of the amount of the second. Before safe levels of usage in particular products can be specified we need estimates of patterns of dietary consumption. It is, furthermore, not enough just to estimate average rates of consumption, because those who need particular protection are those who eat an extreme diet, one which is especially full of additives. The estimation of average and extreme dietary intakes of additives is a particularly complicated business.

Only in the report of the Food Additives and Contaminants Committee on colouring additives, published in 1979, does the British government provide any quantitative estimates of both the amounts of particular additives consumed in specific types of products and overall. The methods by which these calculations were made are open to several objections; in particular, it has been argued that their supposedly 'extreme' diets substantially underestimate the amounts which some people, especially children, may be eating. Only if we know what people eat and how much of each additive is in each product, can we estimate whether or not their intake equals or exceeds the ADI. Unfortunately we do not have the requisite data, and so that is one set of calculations which we cannot make. It may well be that ADIs are of little value, and whatever value they may have is no use at all if we cannot also estimate actual rates of intake for particular chemicals.

Even if the use of ADIs and SFs can be of some value, it is only so where we can assume that there is some threshold level below which the chemical can safely be used. There is heated debate as to whether or not it is possible to specify a threshold for the safe exposure to chemicals which are known to cause cancer in animals when given to them at high levels. There are probably a few chemicals for which thresholds may possibly be specified, but in general this is not the case. For most chemicals known or suspected to cause cancer to animals or humans we cannot assume the existence of any safe threshold level, and so the device of the ADI cannot be applied.

In an attempt to deal with this problem a variety of approaches have been suggested, both by those who seek to use potentially carcinogenic chemicals, and also by those, particularly in North America, who are employed by the government to provide technical advice for regulatory decisions. Enormous efforts have been devoted to trying to develop mathematical models with which to estimate risks to humans from low-level exposure to chemicals which are known to cause cancer to animals when fed to them at high levels. The dose levels which are employed in animal cancer tests are often very high indeed. We want to know the likely effects on humans of the long-term low-dose consumption of those chemicals. The problem is made especially complicated by the fact that there are several different possible ways of modelling the extrapolation, which generate widely different results.

The central issue can most easily be presented graphically (Figure 1).

The animal tests provide us with a few points in the high-dose high-response region, and one at the origin (O). The issue is to identify a plausible shape for what is called the dose-response curve. A moderately cautious approach would be to assume a linear slope. The most cautious would involve assuming the curve to be super-linear, but the least cautious approach is to assume a sub-linear curve. Many chemicals will only be of economic value to the food industry if it is allowed to use them at levels that might be acceptable if a sub-linear extrapolation is justified; otherwise the permissible levels would be too low to be valuable to the industry. Often the problem is that the animal data are consistent with any and all of the three approaches, and so the issue cannot be settled empirically or scientifically. As a result, the selection becomes a political rather than a scientific matter.

The crucial aspect of the problem can be illustrated by considering the example of the artificial sweetener saccharin. Mathematical models have been developed to estimate the scale of the hazard to the American population from an average level of consumption of 120 milligrammes a day, which corresponds to a few cans of artificially sweetened soft drink for each person. At high doses saccharin can be shown to cause bladder cancers in laboratory rats. The lowest

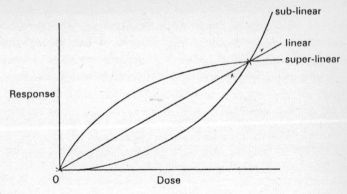

Figure 1

estimate (assuming a sub-linear extrapolation) that has been made of the likely number of cases of human cancer over an average life span from consuming saccharin was 0.22 deaths, while the highest estimate (using a linear extrapolation) was 1,144,000.[2] The difference between them is staggering. If the first estimate is correct then saccharin is a relatively safe chemical. If the other estimate is correct, saccharin would be a major killer. Animal test data are entirely consistent with both estimates, and a toxicological evaluation of saccharin can be no more than an informed political guess.

We can gain a useful perspective on the validity and adequacy of food additive toxicology by comparing it with the extent of our knowledge of the effects of nuclear radiation. Nuclear radiation is a relevant comparison because this is the area of chronic toxicology about which our knowledge is most extensive. Sadly this is a consequence of the use of nuclear weapons on two Japanese cities in 1945, and of the irresponsible exposure of military and civilian personnel to the fall-out of nuclear weapons tests. From these events we have relatively accurate data on the levels of nuclear radiation, and we have relatively detailed information on the subsequent patterns of radiation-related illnesses and death. Unfortunately, the information that we have concerns the effects of high doses of radiation, and that information cannot give us accurate estimates of the hazards from low doses over long periods. Not even in the case of nuclear radiation, where our knowledge is relatively extensive, can we accomplish the requisite extrapolations with any confidence. For all other potentially toxic hazards in our environment, our knowledge is many orders of magnitude worse. As a former Director of the British Industrial Biological Research Association said:

... if you, in the radiation industry, if I may put it like that, cannot define more accurately the lower limits of the dose response curves, so that you are forced into using the straight-line theoretical extrapolation, what hope is there for the chemical industry?[3]

What this means is that toxicology, even at its best, has to rely on linear extrapolation models. Many industrial scientists and executives would prefer to be able to justify the use of sub-linear extrapolations, but they are unable to do so, and consequently find that the levels at which they may use some known carcinogens are more tightly restricted than they would wish. Consumers may wonder that if toxicology is such a poor science, what hope is there for them, let alone for the industry?

Most European countries utilize the device of the ADI in their evaluation and regulation of additives, and they generally set quantitative restrictions on their use. ADIs are set by JECFA, the SCF and the USFDA. That their ADIs frequently do not coincide is a reflection of the intrinsic uncertainties in toxicology, as well as political differences between the various institutions. The USFDA, which is peculiarly given to quantitative analyses, both develops ADIs and utilizes the more complicated mathematical extrapolation techniques. British regulatory institutions assert that they do not use either ADIs or mathematical modelling techniques, unfortunately without providing any specific account of what extrapolatory methods they do use. All they say is that they review the totality of the data and make an overall judgement about the safety of particular chemicals, and the uses to which they may be put. The real point at issue now, though, is not whether one particular method of extrapolation is better than another, but whether, at its present stage of development, toxicological science allows for any consistent and reliable method of extrapolation to humans at all?

There have been persistent warnings in the toxicological literature for at least thirty years that suggest testing chemicals on animals for chronic toxicity does not provide us with any reliable guidance on the risks to, and safety for, humans. These warnings have been almost entirely ignored, and regulatory toxicology has continued as if its foundations had remained unchallenged. This has been possible primarily because the challenges have been confined to the relative obscurity of academic literature. This state of affairs is, however, unlikely to continue.

As early as 1954, before the UK had developed its first positive list, warning signals were already apparent. In a seminal paper, two scientists, Barnes and Denz, carefully reviewed what little evidence was then available on the validity of the experimental and evaluative methods used in the assessment of chronic toxicity. They concluded that there was no evidence that the animal tests could provide a valid basis for conclusions about human toxicology.[4] Even then they could already see that long-term feeding tests are conducted '... as measures of administrative expediency ...' not because they are known to be scientifically valid models of human reality. Since then problems for regulatory toxicologists have only become worse.

In 1961 Litchfield published his conclusions from a complex and careful review of the pharmacology of six drugs in humans, dogs and rats and said that '... many of the most serious side-effects that can result when a drug is given to man were not predictable from observations on dogs or rats.'[5] Since then toxicologists have developed a great deal more precision and have improved the standards by which animal studies are conducted, but they have failed to provide satisfactory replies to the long-standing fundamental criticisms of animal tests. Moreover, the problems have become even more severe, for, as Stevenson has pointed out, '... the 1970s may well be characterized as an era in which toxicology created more problems than it has solved.'[6]

In 1978 a paper of enormous potential significance was published, although it has subsequently been almost entirely ignored. An

American scientist, S. H. Kon, argued that the way in which the chronic toxicity of food additives is assessed is very likely severely to underestimate their risks.[7] The issue to which Kon addresses himself is the vital question of the assumptions which are being made about the shape which we can expect the dose-response curve to possess. Kon spells out an assumption which all industrial and regulatory toxicologists are making but which invariably goes unspoken, unacknowledged and unconsidered. The rule which is being assumed is that: 'between two unequal chronic low doses, the lower one must not be more toxic than the higher'. If the rule is sound then we could conclude that any chronic dose which is smaller than one deemed safe is inevitably safe too and, in fact, safer. This rule is fundamental to the interpretation of a great deal of toxicological data, but Kon argues that this assumption is unjustified, unjustifiable and demonstrably false.

The assumption which Kon is questioning is that the dose-response curve always points towards the origin. Kon argues that it need not do so, and often does not do so, and should not be expected to. The problem can be graphically illustrated.

Earlier in this chapter, three different-shaped dose-response curves were discussed. Although they have different shapes and slopes, all of them slope towards the origin over their entire range. The two shapes represented in Figure 2 differ from each other, and from the previous three. The curve on the left exhibits a maximum and that on the right a minimum point. What they have in common is that they both include a component in which the slope does not point towards the origin. Kon calls these two 'non-monotonic' curves by which he means that the 'tone' of these two curves changes.

It is difficult to exaggerate the importance of the question of the shape of the dose-response curve. If we can assume that all dose-response curves always tend monotonically towards the origin then we can be certain that a lower dose of any chemical will invariably produce less of an effect. Therefore if we can identify a genuine no-effect level in any species then we can specify an acceptable daily intake for that species. Kon is not directly arguing that the results of animal tests underestimate the chronic toxicities for human, but rather he is suggesting that standard methods of evaluating the results of animal tests are likely to be underestimating the low-dose chronic risks to those animals. Obviously, if we are underestimating the risks to the animals then we will consequently underestimate the risks to humans, but Kon's primary concern is with intra-species extrapolation, rather than with extrapolation between species.

The central implication of Kon's argument is that in some circumstances lower doses may be more poisonous than higher doses. He is not just saying that this might happen, he documents seventy cases, drawn from fifteen different additives, in which this has occurred in animal tests. (These additives include: Black PN (E 151), Carmoisine (E 122), Erythrosine (E 127), Propyl Gallate (E 310) and Sunset Yellow (E 110).) But Kon provides us with even more than this. Not only does he identify empirical evidence that the response can rise as the dose falls, he also provides an explanation of the kinds of

Figure 2

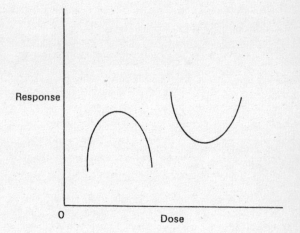

mechanisms, particularly involving the effect of catalysts, which can account for the phenomenon.

Kon's argument is vital because in practice where we do have evidence that lower doses are more toxic than higher doses the evidence is being systematically disregarded. The standard expression which is used to dismiss such evidence as unreal and irrelevant is to say that it is 'not a true dose-response relationship', and therefore that it can be disregarded and dismissed. Such effects are said not to be 'dose-related'. Kon is saying not merely that this evidence should not be dismissed, but that this is the most important kind of evidence. Since toxicologists ignore or dismiss dose-response curves which fail to conform to their prejudices, the chronic toxicity of some chemicals is being systematically underestimated.

Kon concludes his paper with a call for an open and uninhibited discussion of the issues. Since its publication, the silence which this paper has provoked has been entirely deafening. A handful of scholars noticed it, but nobody has either replied to it, or taken its implications on board. The responsibility for doing so lies with the professions of toxicology and pharmacology, and with regulatory authorities, but none of them seem to have noticed.

While Kon has raised some fundamental questions about extrapolations from high doses to lower doses, David Salsburg has raised some equally fundamental questions concerning the extrapolations from laboratory animals to human beings. Salsburg's argument implies not just that no consistent rule of extrapolation is being used, but the more profound claim that, with the current state of knowledge, extrapolations from animals to humans cannot validly and consistently be made.

As recently as the start of 1983 we had to say that we simply had no estimates of the extent of the correlation between the result of animal tests and human toxicity. In that year Salsburg published the first comprehensive quantitative estimate of the validity of long-term feeding studies of chemicals with rodents as a test for carcinogenicity.[8] Salsburg has not only reiterated the point that the validity of animal bioassays for human beings has never been established, but rather he provided the first quantitative evidence that the tests are just not valid! Salsburg compared the results of a sample of long-term feeding studies, and other chronic carcinogenicity animal studies, with what is known of the carcinogenic toxicity of the tested chemicals to humans. His central conclusion was that only about 37 per cent of the chemicals which are known to be carcinogenic to humans have been found to cause cancer in animals in long-term feeding studies. In other words, when it comes to predicting the toxicity of chemicals known to be carcinogenic to humans the tests are wrong more often than they are right!

Salsburg's work has provoked some anxiety amongst professional toxicologists and within regulatory authorities, but perhaps not as much anxiety as the argument warrants. If what he is saying is correct then its significance is very profound. Salsburg is in effect arguing that animal tests are not just a poor guide to human toxicity, but that they are actually misleading. If we tried to determine the carcinogenicity of each known chemical carcinogen by tossing a coin we can be confident that we would get the right answer on average 50 per cent of the time. Salsburg is arguing that the animal tests get it right less often than the tossed coin. If the tests are wrong more often than they are right then they are not just inaccurate but systematically misleading.

Ideally what we want from our toxicological models is information which enables us to identify as precisely as possible those chemicals and doses that are harmful and those that are safe. If a model misleads us into thinking something is safe when it is in fact toxic then that is called a 'false negative', whereas if the test suggests that something is harmful when in fact it is safe we call that a 'false positive'. Obviously, what we want from toxicology is the lowest possible rates of false positives and negatives. Salsburg is saying therefore that currently the rate of false negatives for known human carcinogens is unacceptably high.

Salsburg's work has provoked a modest controversy, and his figures have been challenged. The most optimistic reply to Salsburg has estimated that the animal feeding studies correctly identify known human carcinogens not 37 but 75 per cent of the time.[9] We do not yet have a secure basis for choosing between those two figures.

We can say that a correlation of 75 per cent is better science than one of only 37 per cent, but even so, at best it is still a very poor basis for regulatory policy. It implies that the best we can hope for is that the animal tests only fail to detect human carcinogens a quarter of the time.

Whatever reassurance that might be, it is still slight by comparison with what we really need. This is because these estimates have been made for that small class of chemicals known from human epidemiology to be human carcinogens. As a result of their having been specifically detected, they have been subject to greater scrutiny by the animal testers than any other group of chemicals. In practice what has happened is that it has often taken repeated experiments with different species and different dosing regimes before animal carcinogenicity has been detected. Most other chemicals have been tested far less thoroughly and rigorously. From this it follows that we can expect that the rate of false negatives for most chemicals which cause cancer in humans, but which are not yet known to be carcinogens, will be significantly worse than for the special class of chemicals already recognized as human carcinogens.

I have already mentioned that, tobacco smoking apart, we do not know the cause of 99 per cent of all human cancers, and so we must assume that there are many chemicals which are human carcinogens as yet unidentified. For these chemicals we must assume that the correlation between animal tests and human toxicology is even poorer than for known carcinogens.

The catalogue of difficulties facing toxicology which have been described so far is bad enough, but unfortunately this is not the end of it. However limited toxicology may be, we can at least hope that the work is conducted competently and honestly. During the late 1970s and into the early 1980s it has been possible to discover that commercial and industrial toxicology has been stained with incompetence and even dishonesty. The largest single component of this scandal concerns a company known as Industrial Bio-Test or IBT.

IBT had grown over a period of twenty-five years to reach a dominant position in industrial toxicology. It was a private company, based near Chicago, that conducted toxicology tests for commercial contracts. At their peak, they were responsible for some 30 per cent of all toxicology testing in the world, and had completed over 22,000 studies. They specialized in pesticides and pharmaceutical products. Evidence which they provided was responsible for the regulatory approval of thousands of chemical products. Their role in food additive testing was relatively slight, but the inadequacies which were revealed in the work of IBT were not confined to that institution nor to any particular class of chemicals.

The scandal first broke in 1976, but it was not until 1983 that the subsequent court cases were concluded and the company closed down. The story was first uncovered through the diligence of a scientist called Adrian Gross who works for the US Environmental Protection Agency (EPA). His anxieties were first provoked when he was conducting a spot check on the company's laboratory notebooks, and he encountered the acronym TBD/TDA. This, it transpired, stood for 'too badly decomposed – technician destroyed animal'.

If an animal dies during the course of a test it should be carefully examined so that the condition of the animal, and the cause of death, can be identified. Instead, at IBT, the corpses of thousands of animals had simply been thrown away. A responsible laboratory would have scrapped the rest of the study and started again from scratch. The people at IBT behaved in an irresponsible and illegal manner. They just obtained a new shipment of animals and replaced the dead ones with new ones. By doing so they were entirely invalidating the study, but they were also concealing their actions. The investigation eventually revealed that the results of thousands of IBT studies were worthless. Not only had animals died, and substitutes been put in their place, but test animals had escaped into the local community, and wild rats and mice had infiltrated the laboratory population. Often, proper observations of the animals and detailed pathological tests were not conducted, and in those cases some of the personnel simply made up fake results, which always appeared to confirm the safety of the substances being tested.

When the investigation into IBT was under way, company executives went to great lengths to conceal the extent of their activities, and it was eventually revealed that they had organized the shredding of hundreds of incriminating documents. At the trial in 1983 which followed the investigation, four senior scientists and company executives were accused of palming off the government with false information as scientific data, and three of them were convicted. The fourth person was granted a mis-trial while he underwent heart surgery, and he may face a re-trial.

Following the revelations in this scandal, leading companies have had to spend a great deal of money to repeat some of the studies which have been shown to be inadequate by the standards which we now demand. It is clear, however, that we do not know how many chemicals were not properly tested and which have not subsequently been retested. In 1983 the EPA estimated that of 1205 crucial studies conducted at IBT, only 214 could be considered valid (by orthodox standards), and by then only 254 studies out of the crucial 991 inadequate studies were being repeated.

Subsequently, the EPA and FDA have reviewed the standards of conduct in numerous commercial, industrial and academic laboratories in the USA. They reluctantly concluded that standards in the rest of the industry were not a great deal better than at IBT. Often what they found was not so much blatant dishonesty, as at IBT, but pathetic incompetence amongst laboratory staff. Indeed Edith Efron claims that the standards of honesty and competence in university laboratories was if anything worse than in the commercial establishments.[10]

The response of the US authorities, and those in most other major countries, has been to specify and enforce standards which define what is called 'good laboratory practice' or 'GLP'. These standards have only been in force for a few years, and it is too soon to tell how consistently they are being observed. More importantly, we are not able to establish retrospectively how often governments have relied on the results of studies which were not properly conducted. By examining the toxicological and regulatory literature we can discover that JECFA and other regulatory agencies had been heavily reliant on (frequently unpublished) data from IBT, but we are not able to establish which chemicals still need to be retested and which tests meet the requisite standards.

The scandal has been almost entirely confined to the USA. It would appear that standards of training, competence, honesty and laboratory practice in Britain and the rest of Europe have been significantly higher than those in the USA. What is clear however is that the results of some dubious American studies have played a role in obtaining regulatory permission for additives in Britain and the rest of the EEC. One of our problems is, however, that we are in no position to know which tests come up to scratch and which do not. For all these reasons, and many more besides, we must conclude that toxicity testing as it is currently conducted cannot be said to give us reliable indicators of toxicity to humans.

It is hardly surprising that the leading American toxicologist Wodicka has said, 'Given all the uncertainties, it may seem that the very laborious and expensive tests for assessing safety may be no better than throwing darts at a board full of numbers.'[11] Or, as Goldberg (the founding director of the British Industrial Biological Research Association) has put it, it is as if the scientists are looking under a street light for a key which they know is lost in the middle of a darkened field – they don't expect to find what they are looking for, but there is no point looking in the right place, because there is no light there at all.[12]

Only once we understand the limitations of toxicology is it possible to appreciate the humour in the following passage which was published in a Christmas edition of *The Lancet*:

It is now quite clear that the best model for human cancer is *Herringus Rufus*. All known chemicals are carcinogenic in this animal – a happy finding since it is now possible to reconcile all the theories of carcinogenesis ... The histology of tumours is identical to that in man but, if necessary, it may be different. Although the red-herring lives only two years, correction factors can be applied such that a herring aged 18 months is equivalent to a human being aged 20–70 years: the incidence of tumours is then found to be the same as, or different from, that in man.[13]

I once asked a senior industrial additive toxicologist just how much we could learn about humans by studying animal-model systems and he replied honestly by saying: 'Your guess is as good as mine.' There are some good reasons for continuing to test chemicals on bacteria and animals, but if a substance is shown not to be toxic to animals this does not prove that it is safe for humans. There is a great deal of work which remains to be done to establish as accurately as possible the human relevance of current test procedures, and then to develop tests which are demonstrably relevant. In the meantime, however, we should no longer continue to act as if we are already in possession of a sound scientific toxicology. It may well be that current methods of testing and evaluating additives are underestimating both the acute and particularly the chronic toxicity of additives, as well as many other types of chemicals.

It would be a mistake to conclude that the testing of additives is pointless or that we can learn nothing from it, but we do require a realistic rather than a complacent assessment of its adequacy. Additive regulatory toxicology is characterized by a profound and pervasive uncertainty that is so severe that it prevents us from making more than a handful of secure judgements about the safety or toxicity of most additives. We do not really know what is and what is not safe. There is a great deal of uncertainty, but not everything is equally uncertain. There are some additives against which there are distinctly more question marks than others. Painful though it is for scientists, politicians and consumers alike, it is important that we all recognize that toxicology is a profoundly unsatisfactory science. The use of animal tests does not enable us reliably to determine what is safe and what is not. In the absence of a definitive science, regulatory decisions are being made on non-scientific grounds, and in Britain these decisions are being made behind closed doors, and under a cloak of official secrecy.

The British government copes with these problems in a unique fashion. It adopts two tactics which other governments do not copy. In the first place the British approach is to avoid as far as possible what we can call 'the numbers game'. They recognize at least some of the uncertainties, and therefore decline (for the most part) to assign quantitative figures either to estimates of acceptably safe levels, or to rates of consumption, or to levels at which particular chemicals can be used. Since, however, the British government appears to be committed to the classic idea that it is the dose level which determines whether some chemical is safe, toxic or therapeutic, we have to conclude that behind closed doors some kind of quantitative and qualitative extrapolations and evaluations are being conducted, but we are never permitted to know how these are being done.

Industrial toxicologists often praise the British government for the 'flexibility and discretion' of its regulatory process. I have always understood that as a euphemism for the suggestion that the British government does not really have any consistent or reliable set of extrapolatory and evaluative procedures. Until the protection of the Official Secrets Act is lifted from the process it is hard for us to do anything other than guess and worry.

By comparing some of the published toxcological literature with government decisions, it appears that the British government is not using any consistent set of standards, and that more often than not it is industry which receives the benefits of the many doubts. Consumers require, and are entitled to demand, that they receive the benefit of the doubt.

Scientists who work for the food industry are often prepared (discreetly) to admit that there is no evidence that animal tests are genuinely relevant to human toxicology, but they continue for commercial and professional reasons to conduct the tests and to present the results as if they were genuinely relevant. They do this for the simple reason that governments insist on animal and bacterial data as a basis for their regulatory decisions. Industry does not really believe that these tests are genuinely informative, but they find it simpler, quicker and cheaper to provide governments with the information which they demand, rather than to argue the scientific point.

In 1982 *Principles and Methods of Toxicology* was published.[14] Excitedly I rushed to obtain a copy of this book, eager to identify the principles which underlie this important science. To my regret, but not to my surprise, I was unable to find any systematic set of principles or theories as a basis for toxicology. It has many methods, and many lacks any coherent theoretical foundations, let alone any adequate and comprehensive basis.

The problems which have been catalogued in this chapter show that toxicology is an unsatisfactory science, but this does not mean that they are insoluble. We can conclude, however, that it will not be possible to solve these problems using the current approach to toxicology. The methods, assumptions and practices of toxicology all need to be reconstructed, and in particular the goals have to be reset. The objective of toxicology testing should not be to give industry inconclusive data with which to satisfy uncritical regulators – it must be, and be seen to be, to protect consumers from toxic hazards.

5 | Who Is Being Protected?

Even though additives are used primarily as a technology for the commercial benefit of industry, and British regulations are decided in secret on the basis of very poor scientific information, it might still just be possible that no harm is being done to the health of consumers. Is there any evidence that additives currently in use in the UK are poisoning consumers?

There are some additives which are permitted in the UK for which there is evidence that they are toxic, and some of these have been banned in other countries. This chapter is a review of some of the evidence of toxic hazards covering a range of different types of additives, and in relation to various kinds of hazards. A comparison between the evidence on hazards and the regulatory decisions in the UK and other countries enables us to establish who is getting the benefit of the doubt behind the closed doors of the British government.

The people who are employed by central and local government to regulate and monitor the use of additives are, for the most part, diligent, overworked and underpaid; the material and political resources available to them are woefully inadequate for the demands placed upon them. The poverty of our regulations is a reflection of the distribution and exercise of political power, not of the personnel involved.

Of all the major industrialized countries, Britain has the weakest set of food additive regulations. There are quite a few additives which the British government permits although they are banned in other countries, although some countries do permit a few additives which are not used in the UK. I am not suggesting that there are no problems with the regulations to be found in other countries, but the problems for British consumers are particularly severe.

The British government admits that the list of additives used here is unusually long but attempts to justify this policy by saying that a longer list of additives ensures that a variety of chemicals are used to accomplish each particular task, and that as a result less of each is used. As a device of imaginative rhetoric this deserves a prize. Not only is there no evidence for this claim, but there is evidence that British people do eat more of many additives than do the inhabitants of other comparable countries. The size of the market for additives in the UK had been consistently 10–15 per cent higher than the average for the EEC as a whole.[1] Moreover, the government has no right to claim that their approach reduces consumption of particular additives since they do not even bother to collect the relevant data.

The conclusion that numerous additives are in use in the UK even though we do not know if they are safe is extremely hard to avoid, especially since a senior industrial toxicologist admitted in 1983 that: 'At the present time about 300 food additives in use await toxicological assessment.'[2]

Evidence of hazards from additives can most readily be divided into two broad categories: first there is some direct epidemiological and clinical evidence of acute hazards (to some people) from a group of additives and, second, there is some indirect evidence from animal and bacterial test systems that there may be some chronic hazards from several different chemicals. These two sets of considerations will be dealt with in turn.

Acute hazards

Toxicology is sufficiently powerful to enable us to identify as toxic chemicals which cause widespread and severe acute problems. Toxicology reaches some of its limitations when we seek to identify acute hazards which are not both severe and widespread. They may be uncommon yet severe, or mild but frequent, or slight and rare. Until recently it was extremely difficult to establish whether or not such problems were occurring. In the last fifteen years, though, quite a lot of progress has been made. One of the most important areas in which our understanding has recently grown concerns the severe and tragic problem of child hyperactivity.

Hyperactivity

The term 'hyperactive' has been in our vocabulary for about fifteen years and is used to describe children whose behaviour is particularly difficult. Hyperactive children are essentially frantic. They have explosive energy, they are entirely uncontrollable, desperately restless and incapable of concentration, rarely sleep, are frequently angry and they make their own and their families' lives thoroughly miserable. Most people who have tried to explain the cause of these problems have looked to some environmental factor to explain its occurrence.

The first suggestion that this problem might be a result of consuming food additives came from an American, Dr Ben Feingold, in the early 1970s. His clinical experience over several years led him to conclude that food additives provided the major single trigger for this condition. Consequently he advocated what has become known as the 'Feingold Diet' as a treatment for this syndrome. This diet essentially involves the elimination of a broad set of common additives which are suspected of provoking behavioural disturbances in children. Feingold and others, such as the British organization Hyperactive Children's Support Group (HACSG), claim to have helped many children and families with this diet, but these claims are challenged by the food industry.

As happens so often, the problem comes down to who is getting the benefit of the doubt. When it comes to testing chemicals for safety the food industry is quite happy to settle for a balance of probabilities when the evidence points in its favour, but when the evidence seems to tell against it, the industry demands absolute certainty before any action is to be taken. There is plenty of anecdotal, circumstantial and clinical evidence that hyperactivity can be caused by food additives, but it is much harder to provide definitive and general proof. The trouble with clinical evidence is that it is rarely conclusive. People can recover from their illness both because of, in spite of or independently of what doctors do. To show that clinical intervention made the crucial difference it is necessary to conduct what is called a 'double-blind' study.

This involves taking two similar groups of patients and treating them, to all appearances, identically. Without the knowledge of either the patients or the treaters, the experimenters must arrange for there to be one crucial difference between the treated group and the control group. In other words, you have to slip some suspect additives into the diets of some hyperactive children, but not into the diets of others, without any of the parties involved knowing, and then have them all evaluated by independent experts.

For a variety of reasons, it is extremely difficult to conduct these studies, and to evaluate them. First it is not obvious that such studies should be permitted – they involve deliberately withholding treatment from some children who can be expected to benefit from it – and second it can be difficult to interpret the results.

One of the main problems concerns the accuracy with which the term 'hyperactive' is used. It has become extremely fashionable, particularly in the American school system over the last twelve years, to label as hyperactive many children whom the teachers find difficult to control. In effect, the diagnostic category has been seriously abused. For institutional reasons, it has become convenient to brand children as hyperactive merely because they reject traditional methods of enforcing discipline in the home and in the schools. Once children have been so labelled it becomes possible to segregate them, and then to subject them to drug and behaviour-modification treatment.

The evidence shows, however, that many, if not all, truly hyperactive children will respond well to an additive-free diet. There were several attempts to conduct double-blind tests of the Feingold hypothesis and diet in the USA during the last ten years. The results of those tests were inconclusive. Some children in the test groups and a few in the control groups did show a significant improvement, but the results were marginal. I believe that this was because insufficient care was taken in the use of the diagnostic label.

British doctors have been altogether more cautious than their American colleagues about the use of the term 'hyperactive'. The majority simply do not use it at all, and consequently many hyperactive children suffer with their condition unrecognized by those who

might help them. Those behavioural and educational psychologists, teachers and doctors who do recognize the condition have used the term with far greater precision than in the USA. As a result, it has been British doctors who have been able to provide the strongest evidence that additives cause hyperactivity. In March 1985 a group from the Institute of Child Health published the first conclusive double-blind evidence to confirm the claims of Feingold and the HACSG.[3] As a result, it is no longer possible for the industrial sceptics to say that there is no good evidence that additives are responsible for hyperactivity. We now have both double-blind study evidence and substantial dossiers of case notes showing that, as Feingold put it, 'We can turn these kids on and off at will simply by regulating their diets.'[4]

Some determined sceptics continue to doubt the reality of the hyperactive syndrome on the grounds that in animal tests it has not been possible to reproduce the symptoms. The lack of correspondence, however, between human reality and animal models merely undermines the credibility of animal tests, and in no way diminishes the reality of human suffering.

Other kinds of acute additive intolerance

Acute problems from additives are not confined to children nor to their behavioural difficulties. There is now an extensive body of evidence to show that a so-far-undetermined portion of the population suffers from a range of problems of allergy and intolerance to some additives.[5] These problems include, but are not exhausted by: eczema, urticaria, rhinitis, asthma and migraine. It is universally conceded that we do not know what proportion of the population suffers from problems of food and additive intolerances, but the Scientific Committee for Food of the EEC estimates that between 0.03 and 0.15 per cent of the population are intolerant of food additives. I am not concerned to offer any alternative numbers, but given that intolerances are hard to detect and diagnose, and that it is only recently that doctors have even begun to look for these intolerances, it would be very surprising if these estimates were not revised upwards frequently and substantially in the years to come.

As to these intolerances there are just two points to be made. First, given that these low numbers are being used as grounds for not restricting the use of the additives implicated, we are entitled to ask: what proportion of the population would have to be demonstrably affected before regulations are tightened? Second, the current official and industrial advice is that those diagnosed as intolerant should avoid the additives of which they are intolerant. They can only do this if they know precisely the composition of every product. It is not good enough for the Royal College of Physicians and the British Nutrition Foundation to propose investigating the possibility of establishing a data bank to which only doctors and dietitians would have access.[6] In the first place, there is no justification for any delay in the construction of this database, and there should be complete declarations of composition for all food products, not just those free of additives suspected of provoking intolerant reactions, and it should be openly accessible, and not just for professionals. Moreover there should be complete additive labelling, not just in retailing, but also in catering and on all 'free samples'.

Many of the chemicals which are suspected of being responsible for provoking hyperactivity and other symptoms of intolerance are synthetic coal-tar colours, but they also include some preservatives, antioxidants, flavours and flavour-enhancers. As readers may wish to have a specific list, one can be found in the appendix.

When confronted by the evidence that some additives provoke intolerant reactions, the position taken by the UK government and the British food industry is to blame the victims. In effect they are saying that the chemicals are safe, but the people are wrong. Their line is that if people exhibit symptoms of intolerance it is the responsibility of the victims to identify precisely which chemicals they cannot tolerate, and they must take responsibility for avoiding those substances.

Sufferers can only protect themselves if they know which chemicals make them ill and which products contain those chemicals. It is extremely difficult for most people to gain both those items of knowledge. Even if we have a diagnosis and know the identity of the offending chemicals they can be very hard to avoid. This is because the labelling regulations are so poor. Those who are both blind and intolerant of additives must be having a dreadful time. There are so many gaps in the requirements for additive labelling that it is very easy to consume additives which you are trying hard to avoid. One way of dealing with this problem would be to require all the additives present to be listed on every product.

The governments of some other countries have adopted an altogether more responsible approach than the British, and they are giving the benefit of the doubt to the sufferers and not to those that cause the suffering. Scandinavia has taken a distinct lead in this area. The Swedish government first took action in 1972, requiring that all products containing the colouring additive tartrazine should declare the fact clearly on the label. Tartrazine (or E 102) was the first of the intolerance-provoking additives to be identified in 1959. Subsequently most of the coal-tar dyes which have been studied have been shown to provoke reactions in some of the sensitive people on whom they have been tested. Since the beginning of 1980 it has been illegal to incorporate any coal-tar dyes in food sold in Sweden (with the trivial exception of caviar and caviar substitute).

The Norwegians have been even more thorough. In 1976 they decided that from 1 January 1978 all synthetic dyes would be banned from all food. Furthermore, since 1980 they have required that all remaining additives appear on the label. In 1979 the government of Finland decided that, as from September 1981, four coal-tar dyes, including tartrazine and amaranth, would be banned, and all others fully labelled. Meanwhile seventeen azo dyes, as they are also known, continue to be permitted and widely used in Britain.

Tartrazine is a bright yellow colouring dye and it is the subject of particular controversy. There is plenty of evidence, since at least 1959, to show that this colour provokes intolerant reactions in sensitive people, and it is also clear that there is a link between sensitivity to tartrazine and intolerance of aspirin. The reason why tartrazine heads the league of criticized additives is because it is one of the most widely used and easily recognized. The relative lack of evidence concerning the other coal-tar dyes could merely be a reflection of the fact that they are used and reviewed less often, but it does not follow that the other colours are safer.

This is important because one of the tactics that industry adopts to resist the introduction of strict regulation is to throw the public a hostage. That is to say, they can pick on one additive or small group of additives and agree to have them banned, as long as they can continue to use all the rest. This is important because there is evidence that the food industry has already identified a couple of hostages which it would be willing to discard in an attempt to protect the rest. One is tartrazine, and another is called erythrosine (or E 127). A few leading British manufacturers and retail chains have already devoted considerable efforts to identifying substitutes for those two colours. However, only a handful of manufacturers who use colourings are ready to do without all the synthetics. It is not that it would be hard for us to do without synthetic colouring, it just that industry has not really bothered to think seriously about the possibility.

One standard diagnostic procedure at the clinic of a leading London hospital dealing with problems of intolerance amongst children is to feed them an additive-free diet for no less than a week, and then to give them a glass of orange squash. They use a product which contains a cocktail of additives including tartrazine, sunset yellow and benzoic acid and is treated by the doctors as a sample of the chemicals most likely to be implicated. It is a very cheap, quick and straightforward test, which many people, especially parents, might care to try for themselves.

Colouring additives have received most of the attention from those who have investigated the causes of hyperactivity and intolerance generally, but colours do not have a monopoly of attention. A few preservatives, antioxidants, flavours and flavour-enhancers have also been criticized, and had their use restricted by some governments. When it comes, however, to chronic hazards, a very wide range of different types of additives has been implicated. Toxicology apart, colouring additives are inevitably the ones most vulnerable to criticism because they are the least necessary and most conspicuous of all the food additives.

Evidence of possible chronic hazards

Sodium nitrite

One additive whose chronic toxicity is particularly problematic is sodium nitrite (or E 250). This chemical, and its first cousin sodium nitrate (E 251), is very widely used despite the fact that there is evidence that it may be hazardous. Sodium nitrate is often used with the nitrite because it tends to decay into sodium nitrite, and therefore can be used to provide a buffer stock of the nitrite. Sodium nitrite is permitted for use in the UK as a preservative and is very widely used. Approximately 150 tonnes of nitrite were used in food in Britain in 1985, and incorporated into some 1 million tonnes of cured meats.

It has long been recognized that at high doses sodium nitrite is acutely poisonous. We can stay alive only as long as our red blood cells can carry oxygen from our lungs to the rest of our bodies. If sodium nitrite contaminates our blood supply it prevents the red blood cells from carrying oxygen. One estimate is that you might kill yourself by eating between 3 and 6 pounds of nitrite-treated ham at one sitting.[7]

Fortunately, it is also the case that a doctor can treat the condition if reached quickly enough. For several biochemical reasons babies and young children are particularly vulnerable to this problem, and cases have been documented where sodium nitrite has caused the death of babies, but they must have consumed doses significantly above those which might be consumed by an average British infant. The British government does recognize the risk and for that reason, since the late 1970s, nitrite has not been permitted in products which are marketed specifically for babies. Unfortunately this does not ensure that babies are fully protected. Foods containing nitrite are not labelled as unsuitable for babies, and the government cannot make up its mind as to what counts as a baby. I know of three different official definitions: one of 3 months, one of 6 months and one of 1 year. (Correspondingly there are three different definitions of a young child, namely up to 1, or 2, or 3 years of age.)

The doubts about the long-term safety of sodium nitrite are

perhaps even more significant. Chronic problems are not caused directly by the nitrite and nitrate but by other substances which are formed through interactions with some of the chemicals in the food. Proteins contain amino acids, and nitrites can combine in both the digestive tract and the cooking process with food amines to produce a range of substances which are known as nitrosamines. Nitrosamines are some of the most notoriously powerful carcinogens, known to be carcinogenic to both animals and humans.

The official justification for the continued use of sodium nitrite is that it has an important role to play as a preservative, preventing bacterial spoilage. It is widely used in bacon, ham, cooked sausages such as salamis, as well as in many processed fish and cheese products. These products are all extremely vulnerable to bacterial contamination and some steps have to be taken to protect consumers from food poisoning. Sodium nitrite is effective in this respect, but that does not by itself explain its continued use. There are other chemicals such as vitamin C (ascorbic acid E 300) which can perform the antibacterial function perfectly adequately, and which do not constitute any known long-term hazards. Sodium nitrite is important to the meat-processing industry because it also functions as a cosmetic. As processed meat gets older, in the absence of nitrite, its colour, flavour and odour will deteriorate. Sodium nitrite is valuable to the industry because it functions to maintain these superficial qualities of the products, so that a product which is 3 or 6 months old can look as good as new. For this reason, sodium nitrite is regulated in some countries as a cosmetic and not as a preservative. Ascorbic acid, which is a perfectly satisfactory preservative, does not have this cosmetic effect and would not prevent products from becoming unsightly, and therefore industry does not consider it to be a satisfactory substitute for sodium nitrite.

In the mid-1970s, and for toxicological reasons, sodium nitrite was banned from meat products in Norway. Industrial pressure forced the Norwegian government reluctantly to reinstate it, but the products in, and levels at, which it may be used are tightly restricted. The incidence of bacterial food poisoning in Norway has fallen consistently over the last ten years, while the incidence in Britain has been rising. This shows that nitrite is neither necessary nor sufficient to protect us from bacterial food poisoning.

When the continued use of sodium nitrite in Britain is defended by industrial scientists they often point to the fact that the levels of sodium nitrite in vegetables and drinking water are often higher than the amounts deliberately introduced as chemical and cosmetic preservatives. Although that may be true, it does not mean that sodium nitrite may safely be used in foods. The nitrites which are incorporated into preserved protein food products are more likely to combine with the amines to produce nitrosamines than those which are present in other foods or in water used for cooking or drinking. It also implies that the major source of dietary nitrites, namely the use of nitrogenating fertilizers in farming, poses at least as great a hazard to consumers as does the use of sodium nitrite as a chemical preservative.

This issue also raises another matter concerning the policy implications of toxicology. Implicitly, or explicitly, industry may argue that since the public is already being subjected to a certain level of hazard, a small addition to that hazard is not significant. But the opposite argument is no less plausible. We might well argue that since people are already exposed to some environmental insult, that is precisely why they need to be carefully protected from any further insults of that type.

This type of dilemma arises in several fields. For example, shortly after the screening in 1983 of a Yorkshire Television documentary on childhood leukaemia in the neighbourhood of the Sellafield nuclear plant a letter was published in *The Times* from a Mr Borron stating that his grandfather had been a doctor in the area from 1906 to 1924, and had then noticed a higher than average rate of cancer in that area. The implication was that elevated cancer rates might have been due to some environmental factor which both predates and is independent of the nuclear reprocessing plant. Assuming that the claim is correct then we might draw at least two different conclusions from it. On the one hand we might conclude that high cancer rates are caused by something other than modern installations, and therefore that the plant is not a problem; alternatively we might conclude that since the community is particularly susceptible to environmental carcinogens they deserve to be especially protected from any further carcinogenic insults.

Antioxidants: BHA and BHT

There are some serious questions about the chronic toxicity of synthetic antioxidants. The two related substances – butylated hydroxyanisole (BHA or E320) and butylated hydroxytoluene (BHT or E321) – have been subject to particular scrutiny and criticism. As so often happens, these two additives were permitted long before anything resembling adequate testing had been conducted. Doubts were raised about the safety of these chemicals quite early on during the testing programme. In 1958 and again in 1963 the British government was recommended to ban BHT by the official committee of experts. On each occasion the industry asked for, and was given, the benefit of the doubt. The government permitted continued use of the chemical as long as further studies were conducted. Industry organized and completed some further defensive research, the results of which served to defuse and deflect some of the criticism, but not eliminate it.

Of all the different additives the safety and toxicity of these antioxidants is the most controversial and problematic. This is because while some people claim that there is evidence that these antioxidants may cause chronic illness, others claim that they may also protect us from some chronic illnesses.

There is clear evidence that BHA and BHT can trigger hyperactive and other intolerance symptoms, but the most serious concerns relate to their putative chronic toxicity and particularly their carcinogenicity. One reason why it is particularly difficult to evaluate these substances is that we know that humans and laboratory animals metabolize them in quite different ways. There is some evidence that they interfere with reproductive and embryonic processes in animals at moderate dose levels. The most complex arguments, however, concern their carcinogenicity. There have been several studies over the years which have shown carcinogenic effects in some species at some doses, while other studies have failed to demonstrate any effect. Similarly some have claimed to have discovered circumstances in which each of these chemicals at some dose levels and in some animals can both inhibit and promote cancers from other toxicants. The most serious evidence that BHA may be carcinogenic comes from one of the best-conducted and most recent studies, carried out in Japan by Professor Ito and his colleagues.

In 1982 the Japanese team reported the results of what is generally conceded to be a particularly well-conducted study of the chronic toxicity of BHA. They showed that BHA, when fed at a level of 2 per cent in the diet, causes cancer to the fore-stomachs of rats of both sexes, and even more so to hamsters.[8] As a result of this work, BHA has effectively been banned from food use in Japan. The Japanese have provided clear evidence that BHA is carcinogenic to laboratory animals yet no further restrictions have been placed on its use in the UK, the USA or in the EEC.

Since the cloak of the Official Secrets Act covers the workings of the advisory committee, we have so far been unable to discover what the British government makes of Ito's results. Fortunately the Scientific Committee for Food (SCF) of the EEC is far more open, and it has revealed the standards and criteria by which Ito's work has been judged.

In 1983, the SCF reported that in response to the Japanese results, the EEC created a working party which met with Ito in Heidelberg in October 1982. The report of the working party is a very revealing document.[9] It shows that the working party identified at least eight possible grounds for avoiding the conclusion that the Japanese had shown that BHA was not sufficiently safe to permit its continued use. It reveals that they were explicitly seeking for reasons to give the benefit of the doubt to the food industry, and there is no evidence that they considered giving the benefit of the doubt to consumers. The report states that BHA is considered indispensable *by the food industry*, but there is no claim that it is vital to consumers. It suggests that BHA may protect consumers against poisoning from oxidized toxins, but there is no evidence that the absence of BHA (for example in some brands of Austrian crisps) has ever constituted any hazard to consumers. In total, eight possible devices for disregarding the Japanese results are attempted, but none are successful. In other words, the working party cannot see anything wrong with the studies, but instead of recommending regulatory restrictions they recommend further research in the hope of finding some adequate reason for overriding the negative implications of the evidence. The report states that although Ito's study gives cause for concern there is no 'short-term risk to health'. But nobody ever said that there was; just the risk of a long-term hazard. The working party called for further studies and more evidence before action is taken in Europe. We can all applaud the call for further research, but this should not be used as a tactic for failing to take any action in the short term, nor as a means for avoiding action in the long term.

The SCF states that '... research ... should be undertaken rapidly so that any doubts on the immediate effects on health of BHA in food can be dissipated'. In other words the SCF are calling not for research to establish whether or not BHA can cause cancer in humans, but merely for defensive research to provide reasons for disregarding the human implications of Ito's work. This is significant because if the purpose of further research were to ensure maximum consumer safety rather than to enable industry to continue to use BHA, the programme of research would be substantially different.

Discussions of the acceptability of BHA and BHT are made even more complicated by the fact that some people claim that these antioxidants may serve to protect consumers. The mechanisms by which consumers may be protected are not well understood, and comprise both theoretical possibilities and extrapolations from animal data. There are theoretical reasons for fearing that oxidized oils may be toxic to humans, although attempts to 'confirm' this in animal studies have been inconclusive. There is, furthermore, some evidence from tests with some species of animals at various doses of both antioxidants and known animal carcinogens, to suggest that these antioxidants can in some circumstances both promote and inhibit the formation of tumours. One reason why both these processes can occur, albeit at different doses and in combination with different carcinogens, is because when they reach the liver the antioxidants stimulate the development of highly reactive enzymes. These enzymes can then play an active role in transforming other chemicals as they pass through the organ. In this way they may contribute to the detoxification of some chemicals which might otherwise do harm, but equally they can contribute to the synthesis of its further chemicals which might themselves be poisonous.

For the most part, it is different people who work on different aspects of the problem, and each group emphasizes the significance of its own results, and underplays the importance of the rest. No one is really in a position to judge the overall net effect on humans of consuming BHA and BHT. We have innumerable fragments of information from different laboratory species, using different doses, and different chemical agents; but we do not know how to extrapolate either from the doses which have been studied to the doses which humans consume, nor how to compare the animal species with each other, let alone how to relate all that data to human experience. The use of these antioxidants therefore amounts to a particularly mysterious and dramatic form of chemical roulette. We may be doing ourselves harm by eating them, or we may be doing harm by eating oils and fats free of synthetic antioxidants. We do not know, and we may never know. Perhaps the safest course of action would be to reduce our overall consumption of oils and fats in line with the best nutritional advice, and to rely on vegetable oils which naturally contain their own antioxidants such as vitamin E.

It would be impossible to detect that there are so many problems and uncertainties about the safety of these antioxidants from a superficial scrutiny of statements from the British government and food industry. They give the impression that they fully understand the position, and know it to be safe, but neither aspect of this apparent confidence is justified. Consumers deserve to be informed about these problems, and we need an entirely new kind of research, namely one which enables us to determine how far we can extrapolate from animals to people, and to improve those extrapolations. In the meantime it should be possible to ensure that it is consumers rather than industry who are receiving the benefit of the many doubts.

Amaranth

The colouring additive about which there is most controversy as regards chronic toxicity is amaranth or E 123. So severe has some of the criticism been that amaranth is now banned in the following countries: the USA, the USSR, Greece, Yugoslavia, Austria, Finland and Norway. It continues to be used, however, in some sixty-three countries, and is one of the most widely used colours in Britain. In 1983 the British food industry incorporated about 50 tonnes of amaranth into our food. In western Europe as a whole, there are perhaps eighteen companies manufacturing amaranth, with a combined annual production of some 300,000 kgs., and amaranth is also used to colour at least 1370 drug products.

Amaranth is used very frequently in sweets and soft drinks; even if you avoid these products, however, but otherwise eat an average British diet you will still not avoid amaranth. Almost every processed food or drink which is coloured red contains amaranth, but so do many other products with related colours. Either on its own, or in combination with other coal-tar dyes, it is used to simulate the presence of caramels, apricots, strawberries, raspberries, cherries, currants, grapes and chocolate.

Until the middle of 1985 amaranth was incorporated into Ribena blackcurrant drink, but was then replaced with grape-skin extract. This is just one sign of the good news that some manufacturers and retailers are turning away from amaranth even before the government requires them to do so. For example, in the summer of 1985 Safeway announced that amaranth, as well as tartrazine and forty-eight other additives, would be removed from their own-label brands, and there are now indications that some other retailers are following suit. It is interesting to recognize that on the industry side it is the retail chains which are taking the initiative, and not the manufacturers. This is a reflection of the fact that the retailers are significantly closer to consumers than are the manufacturers. This fact has implications for the ways in which further changes may come about.

There have been doubts about the safety of amaranth at least since 1938. These became more acute in 1970 when some work in the

Soviet Union suggested that it was carcinogenic. This provoked an extensive discussion particularly in the United States, where the scientists of the FDA concluded and then recommended that amaranth (or Red 2 as it is known in America) should be banned. The US food industry put up an enormous fight to retain the use of amaranth which they eventually lost, and amaranth was banned in the USA in January 1976. It was this decision which was responsible for the extraordinary occasion when the American importers of Smarties had to go to the customs shed and throw away all the red ones, so as to ensure that the rest would be allowed into the USA.

One controversy concerned the validity of the Russian work which had indicted amaranth. One of the crucial studies was conducted by the Russian scientist Andrianova, and showed that 2 per cent of amaranth in the diet of rats caused cancer in thirteen out of a test group of fifteen rats. Other Russian studies indicated that it caused birth defects, still births, sterility and early foetal deaths in rats at relatively low dose levels. Much of this Russian work was judged to be unreliable by the Western toxicological establishment because the control group of animals exhibited a very low background level of cancers. For reasons which have already been explained, a judgement about the toxicity of a chemical is based on the difference between the rates of illness amongst animals in the test groups, with those in the control. For any given rate of illness in the test group, the apparent toxicity of the test substance is an inverse function of the rate of illness in the animals in the control group. The unusual feature of the Russian tests was that the cancer rate amongst the controls was so low, and this has always been used in the West as grounds for dismissing the Russian evidence. It is nowadays very rare for animals in control groups to be free of all cancers, but it does occasionally occur. Chapter 4 explained some of the reasons why there are now higher background cancer rates in animals than were common twenty years ago. These might explain both why the Russian studies had low background rates and why these are now so unusual. But if that explanation were correct then it would undercut the grounds for dismissing some of the evidence against amaranth.

During the 1970s, and at the height of the American controversy about amaranth, the FDA conducted its own series of tests to try to see whether they could confirm or refute the Russian findings. For example they investigated the effects of amaranth on poultry chicks. This work readily suggested that the chemical produced defects in the chick embryos even at very low doses. The evidence was relatively complicated because the dose-response curve showed a 'see-saw' effect. As one of the scientists put it: 'From a certain mid-point, the more you gave of the dye and the less, the more the chicks died.'[10] These results were then dismissed by industrial scientists because they did not conform to what most, but not all, toxicologists like to call 'a true dose-response relationship'. Data such as this cannot simply be disregarded as unreliable; rather it should be investigated further. This did not happen, and to this day has not happened. The issue therefore remains unresolved.

The Russians claimed that their tests had shown that chemically pure amaranth is carcinogenic. Western toxicologists also challenged that claim by suggesting that the substance tested was not chemically pure, and that the toxic effects may have resulted from the impurities. This was rejected by the Russians, but they never provided foreign scientists with a sample of the dye which they had tested. This remains one of the many sources of uncertainty in this case. Since the Russians promptly banned amaranth they have no incentive to repeat their experiments.

Subsequent work conducted by scientists from both the FDA and the food industry indicated that amaranth causes female rodents to reabsorb some of their foetuses. This is a curious and unpleasant phenomenon. In a healthy animal a fertilized egg implants on the wall of the womb, and the embryo then grows in a placenta connected to its mother by an umbilical cord until it is sufficiently mature to be born. One of the ways in which that process can go wrong is for the foetal tissue to be consumed back into the maternal womb, and for no foetus to develop. The evidence of reabsorption was ignored and dismissed.

One of the reasons given for dismissing the results of that work related to the method by which the test substance was administered. The animals did not receive it as part of an unlimited food supply.

It was introduced into the animals by a technique known as 'gavage'. Gavage administration involves forcing a tube down the throat of the animals and pouring a liquid directly into their digestive tracts. Gavage has the advantage that the scientists can control precisely the dose and the diet of the animals, and can even force them to digest materials which they would not otherwise choose to consume. It would almost undoubtedly be a mistake to rely heavily on this technique but there is one set of circumstances in which it is appropriate, and this is when you are trying to test the effect of chemicals to be used in drinks. Gavage models drinking although it cannot model either eating or inhaling a chemical. Since amaranth is often incorporated into soft drinks, and for that matter into red alcoholic drinks too, gavage is probably no less appropriate as a method of administration than incorporating it into solid food. This therefore appears to be one more example of toxicologists dismissing evidence when it suggests what they do not want to hear. Toxicologists do not object to gavage when it does tell them what they want to hear.

Although amaranth was banned in the USSR, the USA and Norway (to mention but a few), it continues to be used in Britain. Further careful animal studies have been conducted, and amaranth is still used, and its use is still defended. This is not because experiments have shown no observable ill-effects, but because those ill-effects have exhibited an eccentric pattern which is hard to interpret. In other words, the doubts remain, while industry continues to receive the benefit of those doubts.

Chocolate Brown HT

Another coal-tar colouring additive about which there is much dispute is called Chocolate Brown HT, or just Brown HT for short (listed in Britain as additive number 155). This is not a single chemical substance but a (seemingly variable) mixture of several different dyestuffs. This mixture has been in use for at least twenty-five years, but still nobody is really sure either what it is, or what it does. Approximately 20 per cent of this mixture remains unidentified, and uncertainties about its safety persist. Until 1975 no official body was prepared to provide an evaluation of this colour because there just was not enough information about it. In that year, and under pressure from the British, the Scientific Committee for Food (SCF) ascribed to it a temporary ADI, while calling for several further studies. By 1979 the position had progressed sufficiently for the Food Additives and Contaminants Committee (FACC) to promote this additive from the class of those about which we know nothing, to the class of those chemicals about which we know just about enough to allow it to be used temporarily while further research is completed and evaluated. They called especially for studies on the ways in which a variety of species metabolize the dyestuff, and they also demanded the evidence from further studies of the effects on animal reproduction and embryos.

In 1981 the Joint Expert Committee on Food Additives was still not satisfied that they knew enough about the dye to give it unconditional acceptance. Despite the fact that the dye was found to be staining some internal organs they gave it a temporary ADI of 0.25 mg/kg, but that is only one tenth of the level which the SCF chose. Even that decision involved ascribing to it a safety factor of 600, but no explanation is provided of why that curious figure was selected. By 1983 sufficient information had become available for the SCF to give it a full, as opposed to a temporary, ADI on the basis of some data obtained with mice.

The continued use of Brown HT by the British food industry does not reflect well either on British industry or their products. For over twenty years, the British government allowed manufacturers to use an additive which was not demonstrably safe, and which was considered undesirable in most other countries. Other Continental EEC countries avoid the use of Brown HT by requiring that their chocolate products contain adequate amounts of cocoa powder. There is nothing, short of intense industrial pressure, to prevent the British government from agreeing to raise British standards to those of the rest of Europe. If it were possible to conduct an epidemiological review of the effects of the consumption of Brown HT, and if that review had been conducted, then we could have said that the government and industry have been conducting an experiment using consumers as guinea-pigs. But since such an epidemiological study cannot be conducted we can only conclude that the government has merely been irresponsible.

Saccharin

If it didn't keep happening we would be entitled to be surprised at finding an additive which is still used despite the fact that there has been evidence of its toxicity for many years. Saccharin was one of the earliest synthetic chemicals to be marketed as a food additive. It was discovered in 1879, a patent for its commercial manufacture was granted in June 1885, and it was then introduced into the market, initially in Europe, but in the USA too by the turn of the century. The earliest recorded official criticism of the safety of saccharin was published in 1890 by the Commission of the Health Association of France who decided to ban the manufacture and importation of saccharin. In 1898, the German government restricted the use of saccharin, expressly banning it in food and drink. Similar action was then taken in Spain, Portugal and Hungary. The first formal international gathering to condemn saccharin as unsafe met in Brussels in 1908 and recommended that the use of saccharin and similar sweeteners for human consumption should be prohibited.[11]

The main justification for the use of artificial sweeteners has always been a reflection of the numerous problems with sugar. For example, one can become obese by eating, amongst other things, too much sugar, and those suffering from diabetes must restrict their sugar consumption severely. Since obesity and diabetes are considered as medical conditions, artificial sweeteners have a dual status both as potential food additives and as pharmaceutical products. It was under the latter relatively unregulated category, that the use of saccharin came to be permitted and promoted.

In 1907 the redoubtable Dr Harvey Wiley, then head of the Division of Chemistry at the US Department of Agriculture (the forerunner of the FDA), recommended that saccharin should be banned in the USA. President Theodore Roosevelt rejected this recommendation because his doctors prescribed saccharin for his daily personal use, but Wiley's persistence obliged the President to establish an expert panel to review the issue. Roosevelt appointed as the panel's chairman one of the co-discoverers of saccharin, who stood to benefit financially from the commercial exploitation of his patents. Despite this prejudicial preparation, the panel reported that although in their opinion saccharin was safe when consumed at low doses they conceded that at high rates of consumption (over 0.3 gms per day) it would be toxic. In 1912, while Taft was president, the use of saccharin was banned in food and soft drinks, but permitted in chewing tobacco and under prescription.

It was the crises of the two world wars which created the conditions in which saccharin gained general acceptance. In Europe and the USA sugar was in short supply at these times, and under these circumstances governments around the world relaxed their restrictions on saccharin, failing to reimpose them when peace was restored. At the end of the First World War, C. L. Alsberg (Wiley's successor) tried in vain to prevent Monsanto, the major manufacturer of saccharin, from expanding its sales, while at the end of the Second World War Germany permitted the use of saccharin in ice-cream.[12]

The British position has always been more relaxed and permissive than those of other industrialized countries. There have been hardly any restrictions on the use of saccharin in Britain and to this day it is extensively used. Since the late 1950s and 1960s the market for so-called 'low-calorie' products has been growing rapidly. From 1945 to 1985 the consumption of artificially sweetened products has grown more than 200-fold. One estimate is that by 1978, the advertising budget for diet soft drinks in the USA had reached $40 million.[13] A particularly revealing report has estimated that, in the US at least, 6 per cent of the population consume some 70 per cent of all the saccharin.[14]

The central issue of the modern debate about toxic hazards from saccharin dates from the early 1970s and has been focused on the issue of its carcinogenicity. To cut an extremely long story rather short, by the mid-1970s there was clear and consistent evidence from at least four feeding studies in the rat, that saccharin causes bladder cancer. The results of short-term mutagenicity tests have been equivocal, and it may be that saccharin causes cancer by a non-genotoxic route. The results of epidemiological studies have also been equivocal. Some studies of groups suffering from bladder cancer have found the victims to have been systematic users of saccharin-containing products, while other studies have failed to demonstrate any such connection. By 1980, the International Agency for Research on Cancer stated: 'There is sufficient evidence that saccharin alone, given at high doses, produces tumours of the urinary tract in male rats and can promote the action of known carcinogens in the bladder of rats of both sexes; and there is limited evidence of its tumourgenicity in mice.'[15]

One of the crucial rat-feeding studies, although commissioned by the US National Academy of Sciences, was conducted in Canada, and the reaction of the Canadian authorities was prompt and responsible. On 9 March 1977 the Canadian Health Protection Branch announced its decision to ban saccharin from food and drink, and this decision has been thoroughly implemented. On 15 April 1977 the FDA announced a similar proposal to ban saccharin from all processed food, in soft drinks, and as a table-top sweetener. Despite the explicit provisions of the so-called Delany Amendment, which obliges the FDA to ban all known carcinogens from the permitted list of food additives, the proposal to ban saccharin in the US has never been implemented. Primarily at the behest of the American food, drink, and additive industries, a massive and successful campaign was organised to force the FDA to desist from implementing any restrictions on the use of saccharin. In the autumn of 1977 the US Congress passed a temporary moratorium which prevented the FDA from banning saccharin, and this has subsequently been renewed at least three times. Instead of banning it, the US Congress only required that all products containing the chemical must carry a warning label to the effect that: '... the use of this product may be hazardous to your health. This product contains saccharin which has been determined to cause cancer in laboratory animals.' (See, for example, plate on p. 31.)

The British response to the fact that saccharin is a carcinogen has been at best dilatory and at worst irresponsible. The official British position is provided by the 1982 report of the Committee on Toxicity which is included in the FACC Report on Sweeteners in Food.[16] The COT accepted that saccharin is known to be an animal carcinogen, but recommended that new restrictions on the use of saccharin should not be introduced because the epidemiological evidence had not confirmed that saccharin is a bladder carcinogen for humans as well as for rodents.[17] This is a classic case of giving all the benefit of the doubt to industry, rather than to consumers.

The epidemiological and the animal evidence appear to contradict each other. But despite the fact that there is far greater consistency between the results of the different animal tests than between the various epidemiological studies, the COT chose to give credence to epidemiology over the animal studies, without adequate scientific grounds. There are many reasons for interpreting the results of the epidemiological studies cautiously. It is possible both that some of saccharin's victims died before the studies started and so have been missed, and that many of them have not yet exhibited their symptoms, and so have yet to be detected. While saccharin may cause bladder cancers in rats, it might be causing cancer at different sites in humans, and no epidemiological studies have been conducted which might reveal such possible effects.

The COT went no further than demoting saccharin from 'Category A' status to that of 'Category B', which by definition should mean that further studies are required, although the additive may be used in the interim. In this case, the COT did not call for any further studies, but proposed merely to keep its eyes on the toxicological literature. The only contribution to the debate by the FACC is to argue that the food industry needs saccharin, and that it benefits consumers, even in the face of evidence to the contrary. For example, an authoritative report states that '... the use of non-caloric sweeteners and artificially sweetened foods considered as a group do not appear to have any particular value in weight reduction', and a special medical report to the National Academy of Sciences stated that '... there is no essential requirement for the use of saccharin in the medical management of diabetics and obese patients'.[18]

Listing saccharin under Category B has had no practical effect. No new restrictions on the use of saccharin have been introduced in the UK, and the use of saccharin has declined only because the newest competitor, aspartame, has captured some of its market. Although aspartame is substantially more expensive than saccharin, it has the advantage of not leaving a bitter aftertaste, and for this reason is displacing saccharin, especially from diet soft drinks.

In March 1978, the Commission of the EEC made four recommendations concerning the use of aspartame. In effect, the British government has accepted two of these, but rejected the others. Saccharin may not be used in baby foods, and when it is used its presence must always be indicated on the label. The EEC also suggested, however, that labels should include a warning of possible dangers, especially to pregnant women and to young children, and that further quantitative restrictions should be introduced, but these two proposals were rejected by the FACC.

The current position, therefore is that the use of saccharin is barely restricted in the UK and the USA, but banned entirely in Canada, and banned from food and drinks in France, Greece and Portugal where it is sold only in tablet form, as a table-top sweetener. There is no evidence that the UK or the USA are likely to ban this known carcinogen from our foods.

Aspartame

Since 1973, the controversy which has raged around aspartame has exceeded those which have afflicted all other additives. On the face of it, we might expect aspartame to be one of the least problematic chemicals. It is synthesized from a combination of two common, vital and naturally occurring amino acids, namely phenylalanine and aspartic acid. Amino acids are the fundamental constituents of proteins, and aspartame is thought to be digested as a protein. There are, however, two central questions to the aspartame controversy: firstly, has it been tested properly (even by the indifferent standards which currently prevail), and secondly, is aspartame safe?

The production and sale of aspartame is dominated by G. D.

Searle & Co. which owns most of the crucial patents. Searle first petitioned the American government for permission to market aspartame as early as 1973, but it was not until 1981 that the FDA permitted its commercial use, limiting it initially to dry-food products. It was only in 1983 that the FDA finally approved its use in carbonated soft drinks, which is its major market.

In 1982 the Food Additives and Contaminants Committee recommended that the use of aspartame should be permitted in Britain. Aspartame came on to our market, along with several other less competitive artificial sweeteners, in September 1983. At that time there was a political rumpus. According to Andrew Veitch, the medical correspondent of the *Guardian*, the Junior Minister of Agriculture, Mrs Peggy Fenner revealed, in a letter to the Labour MP Nigel Spearing, that commercial pressure had persuaded the British government to introduce regulations permitting these new sweeteners while Parliament was in recess, rather than waiting until the new parliamentary session when MPs would be able to debate the issue.[19]

The major controversy over aspartame has taken place in the USA, starting in 1973 and continuing for at least twelve years, but until 1983 the British press remained ignorant of, or indifferent to, the American debates. It has only been since 1983 that the significance of these important arguments has been appreciated by a handful of British commentators.

Searle first filed a petition with the FDA for permission to market aspartame in 1973, and the FDA proposed to grant permission in 1974. Before the consequences of that decision could be implemented, objections were raised by independent scientists alleging that aspartame causes mental retardation, brain lesions and neuroendocrine disorders.[20] Before these issues could be resolved, a further complex set of objections were raised, the most important of which concerned the fact that some scientists claimed that Searle had failed to conduct their safety tests properly, that at best their work had been negligent, and at worst that it had been fraudulent.

The scandal was first uncovered by scientists from the FDA's drug-control division. Dr Adrian Gross and his colleagues discovered, by examining carefully the laboratory records, that a large proportion of Searle's experimental work was profoundly unreliable. In response to these revelations the FDA established two Special Task Forces; one under the auspices of the Bureau of Drugs, reviewed Searle's safety evaluations of their pharmaceutical products, while the second under the Bureau of Foods, examined aspartame.

The aspartame Task Force had to institute careful reviews of as many as fifteen studies which were judged to be 'pivotal' in the sense of being integral to the approval of aspartame.[21] Their own internal review dealt with just three of these tests. Two concerned the potential embryotoxicity and teratogenicity in both rats and mice, while the third studied the carcinogenic potential to rats of a substance known as DKP (short for diketopiperazine), which is a breakdown product of aspartame. The FDA decided not to rely entirely on its own resources to conduct all the reviews, and put pressure on Searle to oblige them to contract with the US Universities Association for Research and Evaluation in Pathology (UAREP) to review and audit the validity of the remaining twelve sets of tests. Some commentators have argued that the members of the UAREP were not properly qualified to conduct the kind of investigation which was required, and consequently that their eventual conclusions cannot be considered to be reliable.[22]

The results of the research by the Bureau of Foods Task Force make difficult but interesting reading. One of the central charges against Searle was that the conclusions of their tests, as described in the documents submitted to the FDA, failed to reflect accurately the raw data generated in the laboratories. The summaries, it was suggested, underestimated the possible toxicity of the chemical, and overestimated its safety when compared to the raw data. There were, moreover, '... significant deviations from acceptable procedures for conducting non-clinical laboratory studies'.[23] It is especially ironic, therefore, that the Task Force Report seems to reproduce the mistake which it criticizes Searle for making. The conclusions of the Task Force Report fail accurately to reflect the information contained in the body of that report. It states that while these three tests were not properly conducted, and although there were marked differences between raw data and the summaries submitted in the petition to the FDA, these differences: '... were not of such a magnitude that they would significantly alter the conclusions of the studies'.[24] The details of the Task Force Report, however, suggest precisely the opposite conclusion.

The Task Force had difficulty in evaluating the studies, in part because in some cases there just was no raw data with which to compare the supposed results. In other cases, it was impossible to determine which were the real raw results, and which were subsequent revisions or summaries. In some contexts, the Task Force had to rely on information and assumptions provided by Searle employees who had not been involved in the original work. At worst, it was impossible to identify the occasion on which a particular animal had died, for example, as the Report says: 'Observation records indicated that animal A23LM was alive at week 88, dead from week 92 through week 104, alive at week 108, and dead at week 112.'[2] I certainly do not believe in reincarnation, and would not expect that the FDA or the FACC did so either.

When reviewing the test on DKP, the Report lists no fewer than fifty-two major discrepancies in the Searle submission. One of the central problems concerned the quantities of DKP supposedly consumed by the rats. The FDA investigators found no fewer than three separate documents with different specifications for the content and the purity of the test substance, and they were unable to establish precisely which specification, if any, was correct. It was impossible to reconcile the quantity of the chemical requisitioned from stores with the quantities supposedly fed to the animals. There were questions raised as to the extent to which the DKP was uniformly incorporated into the animals' food. There is clear evidence to show that the test substance was not properly ground, and inadequately mixed, so that it might have been possible for the animals to avoid the DKP while eating their food.

The disparity between the substance and the conclusion of the FDA Task Force report is hard to understand. The investigators found so many mistakes which were of such a magnitude, and of such importance, that it would seem that no reliance can be placed on the results of these tests. The authors of the Report's conclusion, however, appear to have decided, perhaps for political reasons, to interpret the evidence 'generously', while the evidence invites or even demands a stricter assessment.

In 1978, the UAREP submitted its 1062–page report, which concluded that the twelve studies they had audited were authentic. Despite the fact that these two reviews had concluded that aspartame had been properly tested, and that the substance is safe, the objectors were still not satisfied, and furthermore a new complex set of objections to the safety of aspartame were introduced. In an attempt to resolve the controversy once and for all, the FDA proposed the establishment of a so-called Public Board Of Inquiry (or PBOI). This was a unique institution; the procedure had never previously been used, and in all probability will not be used again.

The PBOI, which consisted of three academic scientists who were independent of both the FDA and Searle, was used as an alternative to the more usual formal evidential hearings, and was thought by some people to be better suited to dealing with the numerous scientific and technical complexities. The establishment of the Board was announced in June 1979, and it met early in 1980, publishing its conclusions in October 1980. They had two sets of issues on their agenda. On one of the crucial issues their view was that aspartame consumption would not pose an increased risk of brain damage resulting in mental retardation, but on the other vital issue they concluded that the evidence available to them did not rule out the possibility that aspartame could induce brain tumours. Consequently the Board recommended that aspartame should not be permitted for use, and required further testing.

In response, all of the parties, namely G. D. Searle & Co., the Bureau of Foods, and the objectors, filed detailed exceptions to those parts of the Board's conclusions with which they disagreed. Nonetheless, it was the responsibility of the Commissioner of the FDA to make a decision, for the Board's role was merely advisory and not decisive. In July 1981, the Commissioner Arthur Hayes Jr, announced his decision to approve the use of aspartame in food products other that soft drinks. In doing so he made it clear that he disagreed with the PBOI's interpretation of the issue concerning brain tumours. Hayes took the view that the available data were sufficient to persuade him that aspartame does not cause brain tumours in laboratory animals. Subsequently, two of the three members of the Board have revised their own judgement and decided that they now agree with Hayes.

The issue is a rather subtle one; it concerns the way in which the experimental results are interpreted. The results of at least one experiment are very difficult to interpret. The reason for this is that the level of cancers in the concurrent control group of animals was unusually high. As a result, if one compares the results of the test group with concurrent controls then there is no statistically significant increase in cancer rates; whereas if one compares the test group with average historical control groups of the same type of animals, in similar test, then one could conclude that there was a statistically significant increase in cancers.

This touches on a problem which infects large areas of toxicology, and is not confined either to tests for cancer or tests on aspartame. The degree of variability in the background incidence of pathological symptoms in laboratory animals is vast, and poorly understood. In the toxicological literature there is an extensive debate on whether the significant comparisons should be with concurrent controls or with historical averages, and the issue is unresolved, and probably unresolvable. In practice, we can find examples of firms and governments choosing comparisons with whichever groups yield the result which they wish to establish. In this case, Hayes and the FDA choose to accept the comparison between test animals and concurrent control, and in doing so were able to cite other examples of relatively high levels of cancer in animals not receiving test substances. I do not think that we can say who is right or wrong; what we can conclude, however, is that regulatory toxicology is too unreliable and too uncertain to enable us to be confident that the safety of aspartame can be established.

In 1982, Searle petitioned the FDA for permission to use aspartame in carbonated soft drinks. The FDA again reviewed the controversial issues, but reconfirmed its interpretation of the evidence, and accordingly granted permission for this new use. In 1983 James Turner (a lawyer acting for the Community Nutrition Institute,[26] and Dr John Olney (of Washington University, St Louis) again pressed the FDA to reconsider their decision. The FDA refused to do so, and in 1984 these objectors filed an appeal in the United States Court of Appeals to force the FDA to conduct formal hearings. In the autumn of 1985, three Appeal Court judges unanimously decided that the FDA had acted properly, and that the objectors had failed to show that aspartame is unsafe. In spite of these institutional decisions, some scientists remain unconvinced both about the adequacy of these tests, and interpretation of some of the results; and further law suits remain pending in US courts.

Searle's official position is that all their tests have been properly conducted. The company was investigated by a Grand Jury, but no charges have been preferred. When asked publicly, Searle employees insist that all the tests were properly conducted, and that aspartame is safe. In 1984, however, a senior scientist at Searle casually told me that aspartame is one of the most thoroughly tested chemicals only because the company had done the tests sloppily in the first place, and had been obliged therefore to be especially careful in conducting subsequent tests.[27]

A question of vital concern to consumers in this country, therefore, is: did Searle rely on the results of these controversial tests when they obtained permission to market aspartame in the UK? It is difficult to establish the answer to this. We have no way of knowing whether or not the British scientists on the COT and the FACC were adequately informed about the doubts and complications, because the Official Secrets Act protects the process from precisely this kind of scrutiny. The FACC report merely says that their approval of aspartame had been delayed '... because of questions, which have now been answered, about the validity of the supporting toxicological data[28]. But this tells us little; British consumers are entitled to know whether or not the COT and the FACC knew all about the data from these studies, and the doubts surrounding them.

When I asked the Food Additives Division of the Ministry of Agriculture, Fisheries and Food precisely which data had been presented to the COT and the FACC they refused to tell me. They referred me, routinely, to the toxicological summary in the FACC report; but we can look in vain for an answer there. The COT report on aspartame refers to thirteen published documents and one set of secret documents. The crucial information can only be found in the secret documents. It came as no surprise that when I asked the Ministry for the vital information they refused to give it to me. During an interview for BBC Television News, Mrs Peggy Fenner, the Junior Minister of Agriculture, was pressed on the issue of secrecy, and impatiently rejected the suggestion that consumers could rely on the government to provide them with the information to which they are entitled, brusquely insisting that we should: 'Ask industry!'[29] Accordingly I approached Searle. At first they too refused to provide me with the crucial information. I pointed out that if they would not furnish me with the information which I requested, then I would be entitled to make that fact public. Several weeks later I was provided with a comprehensive dossier of information, and an extensive and detailed discussion. The files show that the information presented to both the US and UK governments did included a summary of the results of the three most controversial tests. They were included amongst the 112 different studies which provided the basis for regulatory action. I have been unable to discover, however, the extent to which the British government, and its expert committees, knew about the doubts and the uncertainties. None of the members of the expert committees with whom I have been able to speak, have any recollection of these problems being mentioned.

The consequence of all of these facts is that we cannot be certain that the tests to which aspartame had been subjected are adequate, even by the relatively poor standards or best current practice. By itself, this implies that we cannot be confident that aspartame is safe. This problem is made even more severe by the fact there are some scientists who continue to argue that what we already know about aspartame is sufficient to show that it is unsafe, at least for some consumers.

Two of the most persistent critics have been Professor John Olney and Professor Richard Wurtmann (of the Massachusetts Institute of

Technology). Wurtmann has published a long series of papers reporting the results of his research on the safety of aspartame, in which he has argued that serious problems exist. Apparently, Wurtmann uses aspartame himself, and considers it to be safe in low doses, but is worried about effects of consuming large amounts of aspartame especially in combination with carbohydrates. Wurtmann's research has been primarily concerned with the effects of consuming aspartame on the biochemistry of the brain. He has argued that it may disturb brain functions in a complex variety of ways, which may provoke some severe and acute symptoms. In particular, Wurtmann has argued that he has both theoretical and clinical evidence that very high doses of aspartame can provoke epileptic seizures.[30] I understand from Searle employees that Wurtmann has yet to identify the particular individuals whom he has investigated, and so his claims have yet to be corroborated, but that is a general feature of so much of the controversial evidence in this debate. Olney's research has concentrated, on the other hand, on the possibility that aspartame might cause chronic brain damage especially when consumed in combination with monosodium glutamate, and he too remains dissatisfied about its safety.

It is very difficult to evaluate the relative safety of, and hazards from, sugar and from artificial sweeteners, but it seems probable that aspartame is a bit safer than saccharin, and both are probably safer than sugar, but that is not necessarily sufficient to encourage us all to consume any of these chemicals. Before we do so, many of us may demand more information both from the government and from the manufacturers.

6 | What Is To be Done?

Since this book is nothing if not critical of the food industry, it is perhaps worth making the point that the food industry is not trying to poison its customers, and that it is genuinely in the interests of food companies to provide their customers with what are evidently wholesome and nutritious foods. One problem we have, however, is that the combination of our relative ignorance and the character of market forces means that industry is more interested in the image of safety than in safety itself. It is uneconomic for a manufacturer to offer for sale wholesome food that is mistakenly imagined to be harmful; but a manufacturer can comfortably sell food products that subtly undermine its customers' health just as long as nobody notices that this is happening, or if nobody points it out.

The points of view of industry and of consumers do not necessarily coincide. When industry asks, 'Can this chemical safely be used and sold?' what this means for them is: 'Is it safe for us?' This almost amounts to asking the question: 'Will consumers drop dead holding the packet?' As far as their commercial interests are concerned, firms only need to be sure that their products will not do any harm which could be traced back to their products. Consumers want to know: 'Is this safe for my family and me?' but that question presupposes a far more distant time horizon, and a view in a quite different direction to that seen from the industrial point of view. Consumers want to be sure that the foods they eat and the ingredients they contain will do them no harm, irrespective of whether or not that harm can be traced back to its material cause.

This is the context in which both consumers and producers look to the government to protect their interests, and governments have to make a complex set of judgements about how to accommodate the conflicts with which they must deal. The role of the government is crucial.

22—OCTOBER 18, 1978—GUARDIAN (USA)

The question which faces us now is not whether the regulatory system will change, but how it will change. The food industry is strenuously but discreetly campaigning for changes in regulatory policy. Industry is pressing for a relaxation of standards and policies, while making the point that the current regulatory system is unstable and unsustainable.

It is unstable for a complex set of reasons including scientific, technical, economic and political ones. The analytical technology which enables scientists to detect the presence of chemicals is advancing rapidly. The technology is spectacularly sensitive. Unfortunately the science does not match the machinery because almost invariably we do not know what the quantitative data mean in biochemical and toxicological terms. As the detection technology improves it becomes rapidly easier to identify observable toxic effects in laboratory animals, and this irrevocably drives downwards the 'No Effect Levels' in animal studies, and thus the 'Acceptable Daily Intakes' which regulators can accept.

For these and other reasons regulatory bodies are constantly having to upgrade their requirements for toxicology testing. Whenever a new test is devised it fails to replace any of the former ones, and merely supplements them. As a result it is becoming rapidly more expensive to test chemicals, and only where there appears to be a substantial and reliable market will companies choose to invest in the safety tests. This regulatory environment does not suit the industry. As a result, although currently the food and chemical industries co-operate with the government by participating in the regulatory system, unless changes occur within the next few years the extent of that co-operation may well decline.

The direction in which industry appears to wish to change policies and regulations is by weakening them, whereas the arguments in this book indicate that consumers require a far more rigorous policy and far stricter standards. Several times in this century, when faced with a proposal to tighten additive regulations, the industry has argued that any change would be disastrous for consumers. For example, the milk-processing industry argued that they could not survive without using formaldehyde as a preservative, and the bread-making industry argued that they could not survive without a processing aid called agene. In the event, both industries survived the transition relatively easily. The food industry will strenuously resist substantial improvements in regulatory policy, but this would be more an expression of their anxiety rather than their self-interest. On average, the British food industry could benefit commercially from raised standards, at least if they recognized the opportunities these would offer. A few companies might suffer if standards were raised, but regulatory policy should protect consumers and not second-rate companies.

One particular reason why there is a need to tighten regulations is because now it is faced with the realization that consumers are uneasy about food processing and additives, the food industry proposes to develop new process technologies which are more subtle and discreet than those currently in use. The food industry is not keen to process food less, rather it proposes to make its processing less obvious.

All too often, even when people recognize that there is a need for reform, they focus their proposals too narrowly at the consumer end of the food chain. Improvements in additive regulations cannot be brought about solely by consumers changing their shopping habits. Shopping, cooking and eating must be seen as the final steps in a long process, and if reform is required all the links in the food chain have to be involved.

The analysis in Chapter 1 implies that agricultural policies and farming practices have to bear some of the responsibility for some of the undesirable aspects of our food supply. Overproduction stimulates ambitions and opportunities for the gratuitous processing of products. The foods which are overproduced such as meat, dairy products, fat, sugar and alcohol are most often implicated in causing chronic ill health. Additives often play a vital role in marketing these commodities.

The government holds the primary responsibility for bringing about the changes which consumers want and deserve. Industry will not spontaneously regulate itself, and consumers are individually almost powerless. The current British position is that we already have sufficient legislation in place to empower our government to provide consumers with the regulatory system which they deserve. The problem is not in the legislation, but in the policies and the actions.

The basic need and right of consumers is for adequate information. Only the government has the authority to ensure that consumers are adequately informed. This power is not yet being exercised – far from it. The British government is currently colluding with the industry in keeping consumers in undeserved ignorance. This law should be changed so that all information submitted to the government as a basis for a regulatory judgement is publicly available. That rule is specified and applied in several countries. The US Freedom of Information Act is not a perfect instrument but it empowers consumers to have access to a great deal of, if not all, the relevant information. That Act is not always properly enforced, but at least it exists, and is doing some good in America.

America is one amongst several countries where comprehensive food labelling is required. In Poland, for example, it is illegal to sell any food product unless there is a full declaration of the recipe to the state and the public. Consumers' interests would be directly served if the government of this country were to establish a central national register of the quantitative composition of all food products, with the register being publicly accessible.

There have been at least five empirical studies in as many years which have all shown what we knew already, namely that consumers want more information about food additives.[1] There have been none to indicate the contrary.

If regulatory policy and additive standards in the UK are to improve, we shall require institutional reforms. The current position, with responsibility split between the Ministry of Agriculture Fisheries and Food (MAFF) and the Department of Health and Social Security (DHSS) is unsatisfactory in several respects. Nine times out of ten, when I telephone a civil servant in either department with an inquiry about additives, I am referred to someone else in the other department; and I am already in a position to know which question to take to whom.

One proposal is to locate additive regulations entirely in the hands of the DHSS and take this function away from the MAFF. This might have the advantage of making consumer health the sole consideration guiding policy and regulation. We cannot expect MAFF to conform to that requirement because, as Lesley Yeomans of the Consumers' Association has pointed out, MAFF is a department which sees itself as a sponsor of farming and manufacturing interests, and not primarily responsible to or for consumers.[2]

The difficulty with the proposal to locate the responsibility for regulating additives solely in the DHSS is that it absolves MAFF from responsibility for consumer health and safety, and that would be undesirable. Ideally, and in the long run, we need a Ministry of Food with a responsibility to consumers for the entire food chain. The standards of living of those in the rural areas, and particularly those engaged in agriculture, should be dealt with as part of social and regional policy. Consumer diet and health should not be subordinated to the commercial interests of either farmers or food manufacturers.

In practice, if we are to achieve significant improvements in food-additive regulations then there will have to be profound changes to both MAFF and the DHSS, and to their interrelationship. There is a clear need to reform the organization and composition of the advisory committee system. At least two possible approaches could be adopted. The first would involve recognizing the conflicts of interest and institutionalizing them into the advisory system. Thus we could have an advisory committee consisting of three groups of equal size representing consumers, industry and the government. This would have the advantage that consumers would be properly represented and the difference between their interests and those of manufacturers would not be concealed. Furthermore, it would place the government directly and explicitly in the position of having to mediate between them.

One of the main objections to this approach, however, is that it would still not guarantee that the government would always provide consumer protection rather than industrial promotion. Regulatory capture would remain a possibility. If the chaps who spoke for government had been to the same schools, colleges and universities as the chaps from industry, while the consumer representatives were housewives from Hackney, we could not be certain that consumers would receive the benefits of all the doubts.

A preferable way of reaching recommendations for regulatory policy would be to appoint a panel of able but non-expert lay people and empower them to take evidence from anyone they choose. Under this arrangement, experts would give their opinions and their advice to those with the responsibility for policy, but would not themselves be allowed to be responsible for policy. Even with these arrangements it would not be possible to guarantee that regulatory capture could not occur, but it would make it significantly less likely. The primary means for ensuring that industry does not exercise undue influence over the agency set to regulate it, is for the agency to conduct all its business in public. Freedom of information is the key to improved regulatory policy. The application of the Official Secrets Act to the regulation of food chemicals in Britain is the basic fault in the current system.

The advisory committee will need policy guidelines which are far more precise than those which now apply. The current advisory committee's terms of reference are merely to give ministers advice on matters relating to food additives. These need to be strengthened to make explicit the goal of regulation, namely consumer protection. They should be responsible for advising ministers so as to enable them to ensure, as far as possible, that no consumers are exposed to undue risks from food additives. There should also be public procedures, both parliamentary and judicial, under which it would be possible to review the adequacy with which the committees were satisfying their goals and guidelines. This is already the position in the USA. Since the FDA knows that its actions can be publicly scrutinized it has to act cautiously. It may be that the British regulatory authorities have been acting no less cautiously, but as long as their actions remain concealed behind the cloak of official secrecy we cannot be entirely confident.

The evidence provided in Chapter 5 implies that it is too generous an assumption to believe the authorities are cautious. Since the British government continues to permit the use of dubious additives, as well as those about which further toxicological information is required (that is, those in category B lists), we are entitled to conclude that consumers are not getting the benefit of the doubt as they deserve. A prudent policy would have been, and now would be, to ban all coal-tar dyes and all other additives which are not both necessary and demonstrably safe.

A slightly less austere approach would be to divide additives into two broad classes. The first group would be those which provide either consumer protection or significant nutritional benefit, while the second would comprise all the rest. The first group could then be treated far more favourably than the rest. Only for these additives should we compare the possible risks from using them against the possible risks from not using them. It may be that if we had freedom and adequacy of information then consumers would choose not to buy any products containing additives in the second group, and their use would come to a rapid end without their needing to be banned. Alternatively they may continue in widespread and popular use, but with the informed consent of the public. It is not the purpose of this book to tell consumers what to eat or which institutional arrangements should be adopted. The purpose is rather to enable consumers to appreciate that there are choices which need to be made.

There have been a few proposals to introduce taxation to discourage gratuitous processing or reliance on undesirable additives. Three sorts of taxes have been suggested. One proposal is for a tax on extracted nutritional value: a value extracted tax. This means that manufacturers would be penalized for using processes which impair the nutritional value of the foods which pass through their hands. There is also a parallel proposal to tax the use of synthetic substitutes in proportion to the reduced nutritional value of the products into which they are incorporated. Attractive though those suggestions may be in principle, they would be unrealistically bureaucratic in practice.

The same objection could be made to the proposal to tax processing which adds economic but not nutritional value. The purpose of regulatory policy should not be to discourage or prohibit convenience foods as long as they are wholesome. Good food is sufficiently expensive for us not to wish to raise its price with added taxation. If we increase the price of good food then we will merely ensure that the poor continue to eat a significantly worse diet than those of us who are better off.

It has also been proposed that there should be a tax on dubious additives. The obvious reply to this is that if there are doubts about the safety of an additive then it should not be permitted in the first place, and so the idea of taxing its use should not arise.

There is a vital need for a thorough reorganization and redirection of food research in the UK. In 1982 the Advisory Council for Applied Research and Development advised the government to reduce spending on agricultural research, and food-safety research, and spend more to accelerate the rate of change in food-processing technology. The government accepted this advice, and if that decision is not rescinded soon the long-run costs for consumers could be substantial. There is no reason for the British government to subsidize commercial research and development on food-processing technology except where that technology has a contribution to make by improving consumer safety. As to the rest, private market mechanisms are sufficient. There are many monetarists in farming and food processing who insist that government support for anyone but themselves is covert socialism.

As part of that decision, and as part of the overall reduction in funding for public-sector research and development, the amount of money which the government spends on food-safety research has declined in real terms. The real need, however, is for a substantial increase in spending on research into food safety, and for a new direction in research policy. If this has to be paid for by reductions in other areas of research then the massive surpluses in food production imply that we could readily afford less research and development on improving agricultural productivity.

The regulatory authorities ought to have the power and resources to conduct and commission their own safety tests on commercial products. It may often be appropriate to charge the manufacturers for this work, but in any case the final costs will be borne by consumers either through higher prices or taxes. Most of us would cheerfully spend a fraction more to have higher standards of testing and safety.

The first priority for a new food-safety research policy should be to estimate the extrapolative validity of current toxicological tests and, then (in so far as this is necessary), to develop new tests which are demonstrably valid. No work of this sort is yet being done. There is currently a proposal on MAFF's table from the British Industrial Biological Research Association (BIBRA) for funds to develop new test methods which may produce results which are not significantly less reliable, but cheaper to obtain. No doubt that is what industry wants, but that is not what consumers need.

Regulatory toxicology is in a shambles, and it needs a substantial influx of new personnel, funding and ideas and a reorganization under the direction of those charged with responsibility for consumer protection. The goal of safety-evaluation work should be to determine whether or not chemicals are safe to use, and not, as at present, to provide industry with the data which it requires to negotiate its way through the current regulatory hurdles.

One of the first priorities for any food-safety research programme should be to test what is known as the 'cocktail' effect. Since only rarely do we eat additives singly, they need to be tested both in their common combinations, and in a complex mixture which reflects the overall patterns of usage in our diets. When in the past this has been proposed to MAFF civil servants and industrial toxicologists they have made the point that the results of such tests would be very hard to interpret. That is true, especially given the intellectual poverty of toxicology, but is not a sufficient reason for not conducting the tests.

Another objection to tests on chemical cocktails is that even if the combinations were found to be poisonous it would then be necessary to test each separate chemical individually to see which particular substance was responsible. This argument reflects the atomistic thinking and bureaucratic mentality of those committed to the status quo. It is probable that some chemicals are safe in some combinations, but poisonous in others. The complexities of biochemistry and toxicology should not prevent us from investigating the real world instead of confining our attention to unrealistic models.

One of the most obvious innovations in research which could easily be conducted, and which needs to be conducted, is to monitor the health of food factory workers who handle food additives in bulk. This would provide them with much-needed protection, and would provide extra security for consumers as a whole. When challenged on this question the government says that there is no evidence that food-industry workers are suffering from their exposure to food chemicals. They could say this because nobody had looked for any evidence and nobody was looking. In 1985 however, Melanie Miller, working under the auspices of the London Food Commission and the London Hazards Centre, looked into the problem and promptly found some evidence of hazards.[3] To conduct a proper epidemiological study of the problem would cost quite a lot of money, and take quite a long time, but we now have sufficient evidence to show that such research would produce useful results and should be initiated.

One of the food industry's main objections to any tightening of additive regulations is that it would inhibit research, development and innovation in additives and food products. The promotion of industrial activity, however, is not the goal of regulatory policy. Aside from that, improved regulations would change the direction and character of innovation, but it should not and would not bring it to a halt. Industry always overestimates the difficulties it would face in having to comply with stricter regulations, but there is plenty of evidence to show that regulations can stimulate innovation as much as inhibit it.[4]

Retail companies have an important role to play in improving our food supply. They respond far more readily to consumers' desires and pressure than do manufacturers. The announcement in 1985 by Safeway and Sainsbury's that they would be reducing the quantities and numbers of additives in their own-brand products came about as a result of pressure from their customers, and in response to clear survey evidence that this is what consumers want.

Consumers require an organized pressure group to campaign for improvements in the regulation and use of additives. This organization would have the potential to achieve a great deal. None of the existing consumer organizations or environmental or health pressure groups are yet active on this issue, with the unique exception of the Hyperactive Children's Support Group. Either a new single-issue pressure group, or a coalition of existing groups, is required to work on the problem. Experience suggests that it would probably not be sufficient if the issue of additives were raised on many separate and disconnected agendas. A unified approach will be far more effective. The goal of this organization should be to ensure that British food additive regulations are amongst the best in the world – patently there is no reason to settle for less.

Appendix | Additives Implicated in Hyperactivity and Intolerance

(E) Number	Name
E 102	Tartrazine
E 104	Quinoline Yellow
107	Yellow 2G
E 110	Sunset Yellow FCF
E 120	Cochineal
E 122	Carmoisine
E 123	Amaranth
E 124	Ponceau 4R
E 127	Erythrosine
128	Red 2G
E 132	Indigo Carmine
E 133	Brilliant Blue FCF
E 150	Caramel
E 151	Black PN
154	Brown FK
155	Chocolate Brown HT
E 210	Benzoic Acid
E 211	Sodium Benzoate
E 212	Potassium Benzoate
E 213	Calcium Benzoate
E 214	Ethyl 4-hydroxybenzoate
E 215	Ethyl 4-hydroxybenzoate, sodium salt
E 216	Propyl 4-hydroxybenzoate
E 217	Propyl 4-hydroxybenzoate, sodium salt
E 218	Methyl 4-hydroxybenzoate
E 219	Methyl 4-hydroxybenzoate, sodium salt
E 220	Sulphur dioxide
E 250	Sodium nitrite

E251	Sodium nitrate
E310	Propyl gallate
E311	Octyl gallate
E312	Dodecyl gallate
E320	Butylated hydoxyanisole
E321	Butylated hydoxytoluene
612	Mono-sodium glutamate
622	Mono-potassium glutamate
623	Calcium dihydrogen di-L-glutamate
627	Guanosine 5'-(disodium phosphate)
631	Inosine 5'-(disodium phosphate)
635	Sodium 5'-ribonucleotide

Further information can be obtained from the Hyperactive Children's Support Group. Send a stamped, addressed envelope to The Secretary, HACSG, 59 Meadowside, Angmering, West Sussex, BN17 4BW. Another useful source is the Soil Association pamphlet *Look Again At The Label: Chemical Additives in Food*. Their address is: 86 Colston Street, Bristol BS1 5BB. Readers may wish to consult the forthcoming *Penguin Dictionary of Additives*.

Notes and References

Introduction

1. B. Moore, *The Social Origins of Dictatorship and Democracy*, Peregrine Books, London, 1969, pp. 522-3.
2. D. Zwerdling, 'Death for Dinner', *New York Review of Books*, 21 February 1974, pp. 22–4.

Chapter 1

1. For a useful discussion of the deliberate creation of scarcities, see S. George, *How the Other Half Dies*, Penguin Books, Harmondsworth, 1976 (especially Chapter 6).
2. Advertisement from Custom Food Products Inc., in the April issue of *Food Product Development*, cited by M. Jacobson in *Eater's Digest*, Anchor Doubleday, New York, 1972, p. 10.
3. J. Burns *et al.*, *The Food Industry: Economics and Policies*, Heinemann, London, 1983, p. 2.
4. *Report on the Food Industry and Technology*, Advisory Council for Applied Research and Development, HMSO, London, September 1982.
5. See, for example, G. Brody *et al.*, 'Television Food Commercials Aimed at Children, Family Grocery Shopping, and Mother-child Interactions', *Family Relations*, Vol. 30, July 1981, pp. 435–9.
6. See, for example, M. Haslum *et al.*, 'What Do Our Ten-year Old Children Eat?', *Health Visitor*, Vol. 57, June 1984 (especially p. 179).
7. 'A Boring Diet Is Bad for Industry', *The Economist*, 13 March 1982, p. 75.
8. See, for example, *Labour Research*, December 1981; and *Supermarketing*, 27 April 1984, p. 5.
9. *Food Engineering*, 1971, quoted by J. Thomson, 'Should Modern Food Carry a Government Health Warning?', *World Medicine*, September 1975.

Chapter 2

1. For a useful discussion of sugar and salt see, for example, C. Walker and G. Cannon, *The Food Scandal*, Century, London, 1985 (Parts I–3, I–6, III–7 and III–8).
2. J. Wimberley, *Predicast Industrial Study No. 181: Food Additives*, Predicast, 1980, pp. 9–10, 23, 69.
3. Jack Knights, 'Good Enough to Eat?', Thames Television, 8 October 1985.
4. See, for example, *Environmental Health Report*, Institute of Environmental Health Officers, London, 1985. See also *The Times*, 5 September 1985, p. 3.
5. G. A. H. Elton, 'Allocation of Priorities – Where Do the Real Risks Lie?', in *Food Toxicology: Real or Imaginary Problems?*, G. Gibson and R. Walker (eds), Taylor and Francis, London, 1985, p. 4.
6. See, for example, D. J. Jukes, *Food Legislation in the UK: a Concise Guide*, Butterworths, London, 1984, p. 63.
7. R. Walford, *Maximum Lifespan*, W. W. Norton, New York, 1983.
8. R. Walker, 'Biochemical Aspects of Food Safety', in *Food and Health: Science and Technology*, G. Birch and K. Parker (eds), Applied Science Publishers, London, 1980 (especially p. 178); N. Ito *et al.*, 'Antioxidants: Carcinogenicity and Modifying Activity in Tumorigenesis', *Food Toxicology: Real or Imaginary Problems?*, pp. 181 *et seq*.
9. The actions of antioxidants are also supplemented by a small group of additives called synergists which assist antioxidants in the performance of their functions. There is also a small group of additives called sequestrants which can serve to deactivate metals which could act catalytically to accelerate oxidative degeneration. If oxidation is hazardous then sequestrants too might have a consumer-protective function. It would be more important, however, to determine how the metallic contamination occurred in the first place, and to eliminate the problem at source.
10. See J. Verrett and J. Carper, *Eating May Be Hazardous to Your Health*, Anchor Books, New York, 1975, p. 24.
11. F. Graves *et al.*, July/August 1984. 'How Safe is Your Diet Soft Drink?', *Common Cause*, pp. 25–43.
12. See, for example, F. Clydesdale, 'The Influence on Sensory Perception and Food Choices', in *Developments in Food Colours – 2*, J. Walford (ed.), Elsevier, London, 1974, pp. 75 *et seq.*; see also A. C. Little, 'The

Eyes Have It', *Journal of the American Dietetic Association*, 1980, pp. 688–91 (especially footnotes 15–20).

13. N. Goldenberg, 'Colours – Do We Need Them?', *Why Additives? – The Safety of Food*, British Nutrition Foundation, 1977, pp. 22–3; Food Additives and Contaminants Committee, *Interim Report on the Review of The Colouring Matter in Food Regulations 1973*, 1979, pp. 8–9 (paragraph 13).
14. A. McLean, 'Risk and Benefit in Food and Food Additives', *Proceedings of the Nutrition Society*, Vol. 36, 1977, p. 87.
15. *Meat Trades Journal*, 13 April 1975.
16. A. McLean, *op. cit.*, p. 88.

Chapter 3

1. Food Act 1984, Part 1 Section 1; see, for example, D. J. Jukes, *op. cit.*, p. 4.
2. Food and Drugs Act 1955; see, for example, the account in B. MacGibbon, 'Control of Food and Contaminants in the UK', in *Food Toxicology: Real or Imaginary Problems?*, G. Gibson and R. Walker (eds), pp. 41 *et seq*.
3. Food Additives and Contaminants Committee, *Interim Report on the Review of Colouring Matter in Food Regulations 1973*, HMSO, London, 1979, pp. 8–9 (FAC/REP/29).
4. Professor R. F. Curtis, 24 October 1984, addressing the Society of Chemical Industry, on the work of the Food Advisory Committee.
5. See, for example, *News from Tesco*, 9 January 1985; *Healthy Foods*, Social Surveys (Gallup Poll Ltd), September 1984 (conducted for the Tesco Group of companies; T. Lang *et al.*, *Jam Tomorrow?*, Food Policy Unit, Manchester Polytechnic, 1984, pp. 20–1; *Survey on Attitudes to Food Additives*, Consumers' Association, 24 September 1984.
6. *Look at the Label*, Ministry of Agriculture, Fisheries and Food; free from Publications Unit, Lion House, Willowburn Trading Estate, Alnwick, Northumberland NE66 2PF.
7. *Look Again at the Label*, Soil Association, 86 Colsten Street, Bristol BS15BB (price 50p with S.A.E. or 65p including p. & p.).

Chapter 4

1. See, for example, W. R. Havender, *Of Mice and Men: the Benefits and*

Limitations of Animal Cancer Tests, American Council on Science and Health, Summit, New Jersey, 1984.

2. Dr Christopher Wilkinson, *Proceedings 10th International Congress of Plant Protection*, Vol. 1, 46, 1983, quoted by Ian Graham-Bryce, *Chemistry and Industry*, 17 December 1984, p. 864.

3. D. Conning, *Archives of Toxicology*, Supplement 3, 1980, p. 52.

4. J. M. Barnes and F. A. Denz, 'Experimental Methods Used in Determining Chronic Toxicity: a Critical Review', *Pharmacological Review*, Vol. 6, 1954, pp. 191–242.

5. J. T. Litchfield, 'Forecasting Drug Effects in Man from Studies in Laboratory Animals', *Journal of the American Medical Association*, 8 July 1961, pp. 34 *et seq*.

6. D. E. Stevenson, 'Current Problem in the Choice of Animals for Toxicity Testing', *Journal of Toxicology and Environmental Health*, Vol. 5, 1979, pp. 9–15.

7. S. H. Kon, 'Underestimation of Chronic Toxicities of Food Additives and Chemicals: the Bias of a Phantom Rule', *Medical Hypotheses*, Vol. 4, 1979, pp. 324–39.

8. D. Salsburg, 'The Lifetime Feeding Study in Mice and Rats – an Examination of its Validity as a Bioassay for Human Carcinogens', *Fundamental and Applied Toxicology*, Vol. 3, 1983, pp. 63–7.

9. See the correspondence between J. K. Haseman *et al.* and D. Salsburg in *Fundamental and Applied Toxicology*, May/June 1983, pp. 3a–7a; and later between D. Salsburg and J. K. Haseman in the same journal, 1984, pp. 288–92.

10. E. Efron, *The Apocalyptics: Cancer and the Big Lie: How Politics Controls what we Know About Cancer*, Simon & Schuster, New York, 1984, p. 239.

11. V. O. Wodicka, 'Risk Assessment and Safety Evaluation', in J. V. Rodericks and R. G. Tardiff (eds), *Assessment and Management of Chemical Risks*, American Chemical Society, Washington DC, 1984, p. 143.

12. L. Goldberg, 'Food Toxicology: Time for an Agonizing Reappraisal', in *Food Toxicology: Real or Imaginary Problems?*, G. Gibson and R. Walker (eds), Taylor and Francis, 1985, p. 393.

13. Anon., *The Lancet*, 21 December 1974, p. 1506.

14. A. Wallace Hayes (ed.), *Principles and Methods of Toxicology*, Raven Press, New York, 1982.

Chapter 5

1. Frost and Sullivan, *Food Additives Markets in Europe*, Report No. E-550, September 1982.
2. S. D. Gangolli, 'Toxicological Aspects of Food Safety', *Food Chemistry*, Vol. 11, 1983, p. 341.
3. J. Eggar *et al.*, *Lancet*, 9 March 1985.
4. D. Zwerdling, 'Death for Dinner', *New York Review of Books*, 21 February 1974, p. 22.
5. See the Joint Report of the Royal College of Physicians and the British Nutrition Foundation, 'Food Intolerance and Food Aversion', *Journal of the Royal College of Physicians of London*, Vol. 18, 2, 2 April 1984, pp. 115 *et seq*. The term 'intolerance' is not entirely felicitous because it can be taken to imply that the problems are the fault of the victims. It might be preferable to say that some people are especially vulnerable to additives, but some scientists insist on the term 'intolerance' because of technical difficulties regarding the use of the term 'allergy'.
6. *Ibid.*, p. 120, Recommendation 5.
7. M. Jacobson, *How Sodium Nitrite Can Affect Your Health*, Centre for Science in the Public Interest, Washington DC, 1973, p. 9.
8. N. Ito *et al.*, *op. cit.*, pp. 181 *et seq*.
9. Report of the Scientific Committee for Food, 14th Series, Brussels, 1983, p. 36.
10. Verrett and Carper, *op. cit.* p. 107.
11. D. L. Arnold *et al.*, 'Saccharin: a toxicological and historical perspective', *Toxicology*, Vol. 27, Part 3–4, July–August 1983, pp. 179–256.
12. *Op. cit.*, pp. 180–6.
13. *Op. cit.*, p. 186.
14. National Research Council, *Saccharin: Technical Assessment of Risks and Benefits*, Report No. 1, Committee for a study on saccharin and food safety policy. Prepared for the FDA, Contract no. 223–78–2415, Washington D.C., 197–8; quoted by D. L. Arnold *et al.*, footnote 24.
15. IARC, *Monographs on Evaluation of the Carcinogenic Risk of Chemicals to Humans, Some Non-Nutritive Sweetening Agents*, No. 22, 1980; quoted by D. L. Arnold *et al.*, p. 243.
16. Food Additive and Contaminants Committee, *Report on the Review of Sweeteners in Food*, HMSO, London, 1982.
17. Committee on Toxicity Report, *ibid.*, Appendix II.
18. M. B. McCann *et al.*, *Journal of the American Dietetic Association*,

Vol. 59, 1971, p. 485; and B. Williams, 'Health – Considerations of the Institute of Medicine Committee on Saccharin', in *Academy Forum*, National Academy of Sciences, 1975, p. 163; quoted by D. L. Arnold *et al.*, pp. 240–1.

19. 'Chemical approved after business pressure', The *Guardian*, 28 September 1983, p. 6.
20. *Searle Response to Common Cause article 'How Safe Is Your Diet Soft Drink?'*, G. D. Searle & Co., 1984.
21. See *Memorandum from the Bureau of Foods Task Force*, to H. R. Roberts, of the FDA, Department of health Education and Welfare, 29 September 1977, p. 1.
22. F. Graves, *et al., op. cit.*; Dr A. Gross, personal communication, 24 December 1985.
23. *Memorandum from the Bureau of Foods Task Force*, to H. R. Roberts, of the FDA, Department of health Education and Welfare, 29 September 1977, p.2.
24. *Ibid.*
25. *Op. cit.* p.2.
26. James Turner is also the author of *The Chemical Feast*, Grossman Publishers, New York, 1970.
27. Personal Communication, 16 June 1984. Dr Adrian Gross states, moreover, that the crucial and inadequate tests have never been repeated by any independent scientists.
28. Food Additives and Contaminants Committee, *Report on the Review of Sweeteners in Food*, HMSO, London, 1982, p. 1.
29. BBC 1 Television, *Six O'Clock News*, 28 October 1985.
30. R. J. Wurtmann, 'Aspartame: Possible Effects on Seizure Susceptibility', *The Lancet*, 9 November 1985, p. 1060.

Chapter 6

1. See Chapter 3, note 5.
2. Lesley Yeomans, 'Food Additives: the Balance of Risks and Benefits', Society of Chemical Industry, 15 May 1985.
3. M. Miller, *Danger! Additives at Work*, London Food Commission, London, 1985.
4. Harvey Brooks in 'The Typology of Surprises in Technology, Institutions and Development' says: 'regulations when first propounded are often

technically unrealistic but frequently technology has been more successful in finding a way to meet them than the experts in the affected industries believed possible.' See Chapter 11 of *Sustainable Developments in the Biosphere*, forthcoming.

Index

'acceptable daily intake' (ADI), 87, 88, 89, 92, 125, 140
acesulfame-K, 46–7, 65
acidity regulator, 11, 28
ADI; *see* 'acceptable daily intake'
Advisory Council for Applied Research and Development, 23, 146
allergy (intolerant symptoms), 110–13, 117
'A list', 130
Alsberg, C. L., 127
amaranth (E123), 28, 52, 112, 121–4
Ames, Bruce, 77
animal tests, 38, 53, 78, 79, 81–6, 90, 91, 93, 94, 95, 97, 103, 106, 110, 119, 122–4, 132, 135–6
anti-browning agents, 39
anti-caking agents, 11, 53
antioxidants, 11, 12, 22, 28, 37, 38, 39, 40, 56, 67, 111, 113, 117–20; *see also* BHA (E320) *and* BHT (E321)
artificial sweeteners, 12, 28, 45–6, 60; *see also* aspartame, saccharin *and* acesulfame-K
aspartame, 46–7, 72, 130, 131–8
Assize of Bread, repeal of, 59
azo dyes, 112; *see also* coal-tar dyes

bacterial contamination, 34–5, 36, 115–16

Barnes, J. M., 93
Batchelors, 25
Bayerische Aniline und Soda Fabrik (BASF), 48
benzoic acid, 113
BHA (butylated hydroxyanisole) (E320), 11, 12, 37–8, 56, 117–20; *see also* antioxidants
BHT (butylated hydroxytoluene) (E321), 11, 12, 37–8, 56, 117–20; *see also* antioxidants
Birds Eye, 25
Black PN (E151), 95
British Industrial Biological Research Association (BIBRA), 65, 147
 former director, 92
 founding director, 101
British Nutrition Foundation, 111
Brooke Bond, 25
Bureau of Foods, 134

Cadbury-Schweppes, 61
Canadian Health Protection Branch, 128
cancer, 37, 38, 78, 80, 81, 85–6, 89, 90, 98, 116, 118, 119, 122, 129, 135–6
caramel colours, 53
carcinogenicity tests, 82, 85, 96–7, 98, 117–20, 122, 127, 132, 135–6

carcinogens, 78, 79, 80, 81, 86, 90, 92, 97, 98, 115, 116–17, 117, 119, 129
Carmoisine (E122), 95
'Category A', 129
'Category B', 129, 130, 145
children
  and advertising, 50
  hyperactive, 106, 107–10, 113
Chocolate Brown HT (155), 124–6
citric acid (E330), 12
coal-tar dye, 48, 111, 112, 113, 121–4, 124–6, 145
colourings; see colours
colours, 9, 10, 26, 32, 38, 47–53, 61, 62, 70, 111, 112, 113, 121–4
Commission of the Health Association of France, 126
Committee on Carcinogenicity (COC), 64
Committee on Mutagenicity (COM), 64
Committee on Toxicity (COT), 64, 129, 136
Common Agricultural Policy, 14, 17, 18
Common Market; see EEC
Consumers Association, 143
'cosmetic' additives, 40–41, 62, 115
  colours, 48–53
  textual, 53–4; see also emulsifiers
cyclamates, 46, 60, 72

Delaney Amendment, 128
Denz, F. A., 93
Department of Health and Social Security (DHSS), 64, 65, 143, 144

DKP (diketopiperazine), 132, 133–4

E102; see tartrazine
E110; see sunset yellow
E123; see amaranth
155; see Chocolate Brown HT
E250; see sodium nitrite
E251; see sodium nitrate
E320; see BHA
E321; see BHT
Economist, The, 26
EEC (Common Market), 16, 17, 39, 40, 65, 70, 87, 88, 106, 118, 125, 130
Efron, Edith, 100
emulsifiers, 11, 12, 28, 33, 53–4
Environmental Health Officers, 50
Environmental Protection Agency (EPA), 99, 101
epidemiology, 75–7, 78, 98, 106, 126, 128, 148
Erythrosine (E127), 95, 113
extrusion cooking, 43–4

FAC; see Food Advisory Committee
FACC; see Food Additives and Contaminants Committee
famine, 12
FDA; see Food and Drug Administration
Feingold, Ben, 107, 110
  Diet, 107, 108
Fenner, Peggy, 131, 137
'flavour', 43
'flavoured', 43
flavour-enhancers, 11, 45, 111, 113;

see also monosodium glutamate
flavourings, 12, 22, 28, 33, 41–8, 62, 70, 111, 113
Food Act (1984), 59
Food Advisory Committee (FAC), 61, 64, 65, 66–7, 72
Food and Drug Administration (FDA), 46, 65, 72, 87, 92, 100, 122, 123, 126, 128, 131–6, 145
   aspartame Task Force, 132–4
   Public Board of Inquiry, 134–5
Food and Drugs Act (1938), 59
   (1955), 59
Food Additives and Contaminants Committee (FACC), 50, 61, 63, 72, 125, 129, 130, 133, 136
   Report (1979), 52–3, 63–4, 89
   Report (1982), 129
food chain, 13, 14, 142, 143
Food Engineering, 27
food poisoning, 34–5, 36–7, 115, 115–16
Food Standards Committee (FSC), 61
Freedom of Information Act, 142

gelling agent, 11, 12, 28
General Foods, 25
Goldberg, L., 101
good manufacturing practice, 67–8
Grand Metropolitan, 61
Gross, Adrian, 99, 131

Hayes, Jr, Arthur, 134, 135
Health Education Council, 25
heart disease, 37
House of Commons Select Committee on Agriculture, 72

humectants, 39
Hyperactive Childrens Support Group (HACSG), 107, 110, 151
hyperactivity, 107–10, 111, 113, 117

ICI, 61
Industrial Bio-Test (IBT), 98–100
Institute of Child Health, 110
International Agency for Research on Cancer, 128
Ito, N., 118–19

John West, 25
Joint Expert Committee on Food Additives (JECFA), 65, 86, 87, 88, 92, 100, 125

Kelloggs, 25
Kon, S. H., 94–6

labelling, 67–70, 112, 142
Lancet, The, 101
Liptons, 25
Litchfield, J. T., 93
London Food Commission, 148
London Hazards Centre, 148
Look Again At The Label, 71
Look At The Label, 71

McCain, 25
MacFisheries, 25
MAFF; *see* Ministry of Agriculture, Fisheries and Food
Malthus, Thomas, 15
Marks & Spencer, 49, 61
Mattesons, 25
mechanically recovered meat, 35

Miller, Melanie, 148
Ministry of Agriculture, Fisheries and Food, 39, 44, 50, 60, 61, 62, 64, 67, 72, 136, 143, 144, 147, 148
modified starches, 28, 54
monosodium glutamate, (MSG) (621), 11, 45; *see also* flavour-enhancers
Monsanto, 127
mutagenicity tests, 77–81, 127

National Academy of Sciences, 128, 129
NEL; *see* 'no effect level'
Nestlé, 25
new product development, 23
'no effect level' (NEL), 87, 140
Nutrasweet; *see* aspartame

Official Secrets Act, 64, 103, 118, 136, 144
Olney, John, 136, 137, 138
organoleptic modifiers, 41

Perkin Henry, 48
phenylketonuria, 46–7
polyphosphates, 54–5
preservatives, 12, 28, 33, 36, 37, 39, 40, 50, 56, 62, 66, 70, 79, 111, 113, 114, 115
 definition of, 36
Private Eye, 46
Principles and Methods of Toxicology, 103
Propyl Gallate (E310), 95

Reckitt & Colman, 61

regulation, 17, 36, 58–73, 105, 112, 140–49
Ribena, 121
Rome Food Conference (1974), 15
Roosevelt, Theodore, 126
Rowntree Mackintosh, 23, 26, 61
Royal College of Physicians, 111

saccharin, 12, 28, 45, 90–91, 126–30
'safety factor' (SF), 87, 89
Safeway, 121, 149
Sainsbury's, 141
Sale of Food and Drugs Act (1875), 59
Salsburg, David, 96–7
Scientific Committee for Food (SCF), 65, 87, 88, 92, 110, 118–19, 125
Searle, G. D., & Co., 130–38
Smedley–HP Foods, 61
SF; *see* 'safety factor'
Smith, Adam, 7, 59
sodium nitrate (E251), 114
sodium nitrite (E250), 56, 114–16
stabilizers, 28, 33, 53–4
sulphur dioxide, 50
sunset yellow (E110), 95, 113

tartrazine (E102), 11, 12, 28, 70, 112-13, 121
textural modifiers, 41
Third World, 13, 16, 59
Trading Standards Officers, 50
Truhaut, René, 87

UN Food and Agriculture Organization, 65, 86
Unilever, 23, 25, 61

United States Department of
   Agriculture (USDA), 52, 126
Universities' Association for
   Research and Evaluation in
   Pathology (UAREP), 132, 134

value-added, 20, 27
Veitch, Andrew, 131

Walls, 25
Wiley, Harvey, 52, 126
Wodicka, V. O., 101
World Health Organisation, 65, 86
Wurtmann, Richard, 141–2

Yeomans, Lesley, 143

MORE ABOUT PENGUINS, PELICANS, PEREGRINES AND PUFFINS

For further information about books available from Penguins please write to Dept EP, Penguin Books Ltd, Harmondsworth, Middlesex UB7 0DA.

In the U.S.A.: For a complete list of books available from Penguins in the United States write to Dept DG, Penguin Books, 299 Murray Hill Parkway, East Rutherford, New Jersey 07073.

In Canada: For a complete list of books available from Penguins in Canada write to Penguin Books Canada Ltd, 2801 John Street, Markham, Ontario L3R 1B4.

In Australia: For a complete list of books available from Penguins in Australia write to the Marketing Department, Penguin Books Australia Ltd, P.O. Box 257, Ringwood, Victoria 3134.

In New Zealand: For a complete list of books available from Penguins in New Zealand write to the Marketing Department, Penguin Books (N.Z.) Ltd, Private Bag, Takapuna, Auckland 9.

In India: For a complete list of books available from Penguins in India write to Penguin Overseas Ltd, 706 Eros Apartments, 56 Nehru Place, New Delhi 110019.

PENGUINS ON HEALTH, SPORT AND KEEPING FIT

☐ **Audrey Eyton's F-Plus** £1.95

F-Plan menus for women who lunch at work * snack eaters * keen cooks * freezer-owners * busy dieters using convenience foods * overweight children * drinkers and non-drinkers. 'Your short-cut to the most sensational diet of the century' – *Daily Express*

☐ **The F-Plan Calorie Counter and Fibre Chart**
Audrey Eyton £1.95

An indispensable companion to the F-Plan diet. High-fibre fresh, canned and packaged foods are listed, there's a separate chart for drinks, *plus* a wonderful selection of effortless F-Plan meals.

☐ **The Parents A–Z Penelope Leach** £6.95

From the expert author of *Baby & Child*, this skilled, intelligent and comprehensive guide is by far the best reference book currently available for parents, whether your children are six months, six or sixteen years.

☐ **Woman's Experience of Sex Sheila Kitzinger** £5.95

Fully illustrated with photographs and line drawings, this book explores the riches of women's sexuality at every stage of life. 'A book which any mother could confidently pass on to her daughter – and her partner too' – *Sunday Times*

☐ **Alternative Medicine Andrew Stanway** £3.25

From Acupuncture and Alexander Technique to Macrobiotics and Yoga, Dr Stanway provides an informed and objective guide to thirty-two therapies in alternative medicine.

☐ **Pregnancy Dr Jonathan Scher and Carol Dix** £2.95

Containing the most up-to-date information on pregnancy – the effects of stress, sexual intercourse, drugs, diet, late maternity and genetic disorders – this book is an invaluable and reassuring guide for prospective parents.

PENGUINS ON HEALTH, SPORT AND KEEPING FIT

☐ *Medicines* **Peter Parish** £4.95

Fifth Edition. The usages, dosages and adverse effects of all medicines obtainable on prescription or over the counter are covered in this reference guide, designed for the ordinary reader and everyone in health care.

☐ *Baby & Child* **Penelope Leach** £7.95

A fully illustrated, expert and comprehensive handbook on the first five years of life. 'It stands head and shoulders above anything else available at the moment' – Mary Kenny in the *Spectator*

☐ *Vogue Natural Health and Beauty*
Bronwen Meredith £7.50

Health foods, yoga, spas, recipes, natural remedies and beauty preparations are all included in this superb, fully illustrated guide and companion to the bestselling *Vogue Body and Beauty Book*.

☐ *Pregnancy and Diet* **Rachel Holme** £1.95

With suggested foods, a sample diet-plan of menus and advice on nutrition, this guide shows you how to avoid excessive calories but still eat well and healthily during pregnancy.

☐ *The Penguin Bicycle Handbook* **Rob van der Plas** £4.95

Choosing a bicycle, maintenance, accessories, basic tools, safety, keeping fit – all these subjects and more are covered in this popular, fully illustrated guide to the total bicycle lifestyle.

☐ *Physical Fitness* £1.25

Containing the 5BX 11-minute-a-day plan for men and the XBX 12-minute-a-day plan for women, this book illustrates the famous programmes originally developed by the Royal Canadian Air Force and now used successfully all over the world.

COOKERY AND GARDENING IN PENGUINS

☐ ***Italian Food*** **Elizabeth David** £3.95

'The great book on Italian cooking in English' – Hugh Johnson. 'Certainly the best book we know dealing not only with the food but with the wines of Italy' – *Wine and Food*

☐ ***An Invitation to Indian Cooking*** **Madhur Jaffrey** £2.95

A witty, practical and irresistible handbook on Indian cooking by the presenter of the highly successful BBC television series.

☐ ***The Pastry Book*** **Rosemary Wadey** £2.95

From Beef Wellington to Treacle Tart and Cream-filled Eclairs – here are sweet and savoury recipes for all occasions, plus expert advice that should give you winning results every time.

☐ ***The Cottage Garden*** **Anne Scott-James** £4.95

'Her history is neatly and simply laid out; well-stocked with attractive illustrations' – *The Times*. 'The garden book I have most enjoyed reading in the last few years' – *Observer*

☐ ***Chinese Food*** **Kenneth Lo** £1.95

The popular, step-by-step introduction to the philosophy, practice, menus and delicious recipes of Chinese cooking.

☐ ***The Cuisine of the Rose*** **Mireille Johnston** £5.95

Classic French cooking from Burgundy and Lyonnais, explained with the kind of flair, atmosphere and enthusiasm that only the most exciting cookbooks possess.

COOKERY AND GARDENING IN PENGUINS

☐ ***The Magic Garden*** **Shirley Conran** £3.95

The gardening book for the absolute beginner. 'Whether you have a window box, a patio, an acre or a cabbage patch ... you will enjoy this' – *Daily Express*

☐ ***Mediterranean Cookbook*** **Arabella Boxer** £2.95

A gastronomic grand tour of the region: 'The best book on Mediterranean cookery I have read since Elizabeth David' – *Sunday Express*

☐ ***Favourite Food*** **Josceline Dimbleby** £2.50

These superb recipes, all favourites among Josceline Dimbleby's family and friends, make up 'an inspiration to anyone who likes to be really creative in the kitchen' – Delia Smith

☐ ***The Chocolate Book*** **Helge Rubinstein** £3.95

Part cookery book, part social history, this sumptuous book offers an unbeatable selection of recipes – chocolate cakes, ice-creams, pies, truffles, drinks and savoury dishes galore.

☐ ***Good Healthy Food*** **Gail Duff** £2.50

Mushrooms in Sherry, Lamb with Lemon and Tarragon, Strawberry and Soured Cream Mousse ... You'll find that all the dishes here are tempting and delicious to taste, as well as being healthy to eat.

☐ ***The Adventurous Gardener*** **Christopher Lloyd** £4.95

Prejudiced, delightful and always stimulating, Christopher Lloyd's book is essential reading for everyone who loves gardening. 'Get it and enjoy it' – *Financial Times*

A CHOICE OF PELICANS AND PEREGRINES

☐ **The Knight, the Lady and the Priest**
   Georges Duby £6.95

The acclaimed study of the making of modern marriage in medieval France. 'He has traced this story – sometimes amusing, often horrifying, always startling – in a series of brilliant vignettes' – *Observer*

☐ **The Limits of Soviet Power** **Jonathan Steele** £3.95

The Kremlin's foreign policy – Brezhnev to Chernenko, is discussed in this informed, informative 'wholly invaluable and extraordinarily timely study' – *Guardian*

☐ **Understanding Organizations** **Charles B. Handy** £4.95

Third Edition. Designed as a practical source-book for managers, this Pelican looks at the concepts, key issues and current fashions in tackling organizational problems.

☐ **The Pelican Freud Library: Volume 12** £5.95

Containing the major essays: *Civilization, Society and Religion, Group Psychology* and *Civilization and Its Discontents*; plus other works.

☐ **Windows on the Mind** **Erich Harth** £4.95

Is there a physical explanation for the various phenomena that we call 'mind'? Professor Harth takes in age-old philosophers as well as the latest neuroscientific theories in his masterly study of memory, perception, free will, selfhood, sensation and other richly controversial fields.

☐ **The Pelican History of the World**
   J. M. Roberts £5.95

'A stupendous achievement . . . This is the unrivalled World History for our day' – A. J. P. Taylor

A CHOICE OF
PELICANS AND PEREGRINES

☐ ***A Question of Economics* Peter Donaldson** £4.95

Twenty key issues – from the City and big business to trades unions – clarified and discussed by Peter Donaldson, author of *10 × Economics* and one of our greatest popularizers of economics.

☐ ***Inside the Inner City* Paul Harrison** £4.95

A report on urban poverty and conflict by the author of *Inside the Third World*. 'A major piece of evidence' – *Sunday Times*. 'A classic: it tells us what it is really like to be poor, and why' – *Time Out*

☐ ***What Philosophy Is* Anthony O'Hear** £4.95

What are human beings? How should people act? How do our thoughts and words relate to reality? Contemporary attitudes to these age-old questions are discussed in this new study, an eloquent and brilliant introduction to philosophy today.

☐ ***The Arabs* Peter Mansfield** £4.95

New Edition. 'Should be studied by anyone who wants to know about the Arab world and how the Arabs have become what they are today' – *Sunday Times*

☐ ***Religion and the Rise of Capitalism*
R. H. Tawney** £3.95

The classic study of religious thought of social and economic issues from the later middle ages to the early eighteenth century.

☐ ***The Mathematical Experience*
Philip J. Davis and Reuben Hersh** £7.95

Not since *Gödel, Escher, Bach* has such an entertaining book been written on the relationship of mathematics to the arts and sciences. 'It deserves to be read by everyone ... an instant classic' – *New Scientist*

A CHOICE OF PENGUINS

☐ ***The Complete Penguin Stereo Record and Cassette Guide***
Greenfield, Layton and March £7.95

A new edition, now including information on compact discs. 'One of the few indispensables on the record collector's bookshelf' – *Gramophone*

☐ ***Selected Letters of Malcolm Lowry***
Edited by Harvey Breit and Margerie Bonner Lowry £5.95

'Lowry emerges from these letters not only as an extremely interesting man, but also a lovable one' – Philip Toynbee

☐ ***The First Day on the Somme***
Martin Middlebrook £3.95

1 July 1916 was the blackest day of slaughter in the history of the British Army. 'The soldiers receive the best service a historian can provide: their story told in their own words' – *Guardian*

☐ ***A Better Class of Person*** **John Osborne** £2.50

The playwright's autobiography, 1929–56. 'Splendidly enjoyable' – John Mortimer. 'One of the best, richest and most bitterly truthful autobiographies that I have ever read' – Melvyn Bragg

☐ ***The Winning Streak*** **Goldsmith and Clutterbuck** £2.95

Marks & Spencer, Saatchi & Saatchi, United Biscuits, GEC ... The UK's top companies reveal their formulas for success, in an important and stimulating book that no British manager can afford to ignore.

☐ ***The First World War*** **A. J. P. Taylor** £4.95

'He manages in some 200 illustrated pages to say almost everything that is important ... A special text ... a remarkable collection of photographs' – *Observer*

A CHOICE OF PENGUINS

☐ ***Man and the Natural World*** **Keith Thomas** £4.95

Changing attitudes in England, 1500–1800. 'An encyclopedic study of man's relationship to animals and plants... a book to read again and again' – Paul Theroux, *Sunday Times* Books of the Year

☐ ***Jean Rhys: Letters 1931–66***
Edited by Francis Wyndham and Diana Melly £4.95

'Eloquent and invaluable... her life emerges, and with it a portrait of an unexpectedly indomitable figure' – Marina Warner in the *Sunday Times*

☐ ***The French Revolution*** **Christopher Hibbert** £4.95

'One of the best accounts of the Revolution that I know... Mr Hibbert is outstanding' – J. H. Plumb in the *Sunday Telegraph*

☐ ***Isak Dinesen*** **Judith Thurman** £4.95

The acclaimed life of Karen Blixen, 'beautiful bride, disappointed wife, radiant lover, bereft and widowed woman, writer, sibyl, Scheherazade, child of Lucifer, Baroness; always a unique human being... an assiduously researched and finely narrated biography' – *Books & Bookmen*

☐ ***The Amateur Naturalist***
Gerald Durrell with Lee Durrell £4.95

'Delight... on every page... packed with authoritative writing, learning without pomposity... it represents a real bargain' – *The Times Educational Supplement*. 'What treats are in store for the average British household' – *Daily Express*

☐ ***When the Wind Blows*** **Raymond Briggs** £2.95

'A visual parable against nuclear war: all the more chilling for being in the form of a strip cartoon' – *Sunday Times*. 'The most eloquent anti-Bomb statement you are likely to read' – *Daily Mail*

PENGUIN REFERENCE BOOKS

☐ ***The Penguin Map of the World*** £2.95

Clear, colourful, crammed with information and fully up-to-date, this is a useful map to stick on your wall at home, at school or in the office.

☐ ***The Penguin Map of Europe*** £2.95

Covers all land eastwards to the Urals, southwards to North Africa and up to Syria, Iraq and Iran * Scale = 1:5,500,000 * 4-colour artwork * Features main roads, railways, oil and gas pipelines, plus extra information including national flags, currencies and populations.

☐ ***The Penguin Map of the British Isles*** £2.95

Including the Orkneys, the Shetlands, the Channel Islands and much of Normandy, this excellent map is ideal for planning routes and touring holidays, or as a study aid.

☐ ***The Penguin Dictionary of Quotations*** £3.95

A treasure-trove of over 12,000 new gems and old favourites, from Aesop and Matthew Arnold to Xenophon and Zola.

☐ ***The Penguin Dictionary of Art and Artists*** £3.95

Fifth Edition. 'A vast amount of information intelligently presented, carefully detailed, abreast of current thought and scholarship and easy to read' – *The Times Literary Supplement*

☐ ***The Penguin Pocket Thesaurus*** £2.50

A pocket-sized version of Roget's classic, and an essential companion for all commuters, crossword addicts, students, journalists and the stuck-for-words.

PENGUIN REFERENCE BOOKS

☐ **The Penguin Dictionary of Troublesome Words** £2.50

A witty, straightforward guide to the pitfalls and hotly disputed issues in standard written English, illustrated with examples and including a glossary of grammatical terms and an appendix on punctuation.

☐ **The Penguin Guide to the Law** £8.95

This acclaimed reference book is designed for everyday use, and forms the most comprehensive handbook ever published on the law as it affects the individual.

☐ **The Penguin Dictionary of Religions** £4.95

The rites, beliefs, gods and holy books of all the major religions throughout the world are covered in this book, which is illustrated with charts, maps and line drawings.

☐ **The Penguin Medical Encyclopedia** £4.95

Covers the body and mind in sickness and in health, including drugs, surgery, history, institutions, medical vocabulary and many other aspects. Second Edition. 'Highly commendable' – *Journal of the Institute of Health Education*

☐ **The Penguin Dictionary of Physical Geography** £4.95

This book discusses all the main terms used, in over 5,000 entries illustrated with diagrams and meticulously cross-referenced.

☐ **Roget's Thesaurus** £3.50

Specially adapted for Penguins, Sue Lloyd's acclaimed new version of Roget's original will help you find the right words for your purposes. 'As normal a part of an intelligent household's library as the Bible, Shakespeare or a dictionary' – *Daily Telegraph*

PENGUIN OMNIBUSES

☐ *Life with Jeeves* **P. G. Wodehouse** £3.95

Containing *Right Ho, Jeeves, The Inimitable Jeeves* and *Very Good, Jeeves!* in which Wodehouse lures us, once again, into the evergreen world of Bertie Wooster, his terrifying Aunt Agatha, his man Jeeves and other eggs, good and bad.

☐ *The Penguin Book of Ghost Stories* £4.95

An anthology to set the spine tingling, including stories by Zola, Kleist, Sir Walter Scott, M. R. James, Elizabeth Bowen and A. S. Byatt.

☐ *The Penguin Book of Horror Stories* £4.95

Including stories by Maupassant, Poe, Gautier, Conan Doyle, L. P. Hartley and Ray Bradbury, in a selection of the most horrifying horror from the eighteenth century to the present day.

☐ *The Penguin Complete Novels of Jane Austen* £5.95

Containing the seven great novels: *Sense and Sensibility, Pride and Prejudice, Mansfield Park, Emma, Northanger Abbey, Persuasion* and *Lady Susan*.

☐ *Perfick, Perfick!* **H. E. Bates** £4.95

The adventures of the irrepressible Larkin family, in four novels: *The Darling Buds of May, A Breath of French Air, When the Green Woods Laugh* and *Oh! To Be in England*.

☐ *Famous Trials*
   Harry Hodge and James H. Hodge £3.95

From Madeleine Smith to Dr Crippen and Lord Haw-Haw, this volume contains the most sensational murder and treason trials, selected by John Mortimer from the classic Penguin Famous Trials series.